Fatigue and Fracture Mechanics of Offshore Structures

Fatigue and Fracture Mechanics of Offshore Structures

Editor

Naveen Patniak

Fatigue and Fracture Mechanics of Offshore Structures
Edited by **Naveen Patniak**

Printed in 2017

ISBN: 978-1-68117-371-9

Library of Congress Control Number: 2015941559

© 2016 by
SCITUS Academics LLC,
616, Corporate Way, Suite 2, 4766,
Valley Cottage, NY 10989

www.scitusacademics.com

Contents

Preface

The tubular welded joints used in the construction of offshore structures can experience millions of variable amplitude load cycles during their service life. Such fatigue loading represents a main cause of degradation in these structures. As a result, fatigue is an important consideration in their design. Fatigue and Fracture Mechanisms of Offshore Structures present novel research and the results of wave-induced stress on the operational life of offshore structures. Increasing oil consumption in the world and scarcity of land-oil resources due to political and economical reasons has caused offshore oil exploration and production to become a growing investigation field in the past six decades. The analysis of structures to use energy deposits and other recourses, or for other purposes, in ocean environments requires a special consideration since environmental and loading conditions offshore are very complicated and contain large uncertainties. Offshore structures are continuously subjected to random ocean waves producing stochastic loads that cause mainly fatigue failure in structural components.

Editor

Degradation in Seawater of Structural Adhesives for Hybrid Fibre-Metal Laminated Materials

Cristina Alia[1], María V. Biezma[2], Paz Pinilla[1], José M. Arenas[1], and Juan C. Suárez[1]

[1]Center for Durability and Structural Integrity of Materials (CISDEM-UPM/CSIC), Universidad Politécnica de Madrid, C/Ronda Valencia, 3-28015 Madrid, Spain

[2]Research Group on Degradation and Corrosion of Materials, Universidad de Cantabria, c/Gamazo, 1-39004 Santander, Spain

ABSTRACT

The adhesives used for applications in marine environments are subject to particular chemical conditions, which are mainly characterised by an elevated chlorine ion content and intermittent wetting/drying cycles, among others. These conditions can limit the use of adhesives due to the degradation processes that they experience. In this work, the chemical degradation of two different polymers, polyurethane and vinylester, was studied in natural seawater under immersion for different periods of time. The diffusion coefficients and concentration profiles of water throughout the thickness of the adhesives were obtained. Microstructural changes in the polymer due to the action of water were observed by SEM, and the chemical degradation of the polymer was monitored with the Fourier transform infrared spectroscopy (FTIR) and differential scanning calorimetry (DSC). The degradation of the mechanical properties of the adhesive was determined by creep tests with Mixed Cantilever Beam (MCB) specimens at different temperatures. After 180 days of immersion of the specimens, it was concluded that the J-integral value (depending on the strain) implies a loss of stiffness of 51% and a decrease in the failure load of 59% for the adhesive tested.

INTRODUCTION

Adhesives play a fundamental role in the manufacture and assembly of structural panels made from hybrid fibre/metal materials, which consist of layers of steel around a glass fibre fabric-reinforced vinylester matrix core [1]. These structural materials are used mainly in marine environments (such as ship hulls and offshore wind farms), in which they will be subjected to high relative humidities, temperatures, and chlorine ion concentrations—among other aggressive agents—over long periods of times. These conditions limit the use of adhesives due to the degradation processes that both the polymeric adhesive systems and the adherents themselves experience, which can lead to a deterioration of the mechanical properties of the assembly and ultimately result in the failure of the joint. In addition, water (liquid or vapour) is one of the most common damaging environmental agents to the durability of adhesive joints [2–4]. The majority of adhesive joints are exposed to water by high relative humidity, making it practically impossible to prevent water from diffusing into the interior of the adhesive. The temperature and humidity act jointly on the mechanical behaviour of the adhesive, synergistically accelerating the degradation processes of the adhesive joint. The mechanical properties of the polymeric materials change over time, especially when they are subjected to variable charge environments for a long period of time, making it extremely important to develop tests that accelerate the degradation process to better understand the deterioration of the joint's mechanical resistance under different environmental conditions [5, 6]. Several studies on the degradation mechanisms of structural adhesives in air have been performed that conclude that the main cause of degradation is photooxidation [7–9]. However, the amount of available information on the degradation mechanisms of adhesives in humid conditions, specifically in a marine environment, is much more limited. Studies performed with epoxy resins [10–12] and polyurethane [13–15] have provided some clarification, but data on vinylester resins in the literature remain sparse [16].

Bowditch [17] described the diverse processes that influence the durability of polymeric adhesives in the presence of seawater; the most important of which were plasticisation, swelling, and hydrolysis. The absorption of water by the adhesive led to resin plasticisation, which reduces the joint strength; they also quantified the critical water content with the goal of relating this content to the adhesive chemistry. Burns et al. [18] determined the bulk module for natural elastomers, neoprenes, and polyurethanes after aging them for up to 2 years in artificial seawater, noting that the properties remained very stable during this period. Rutkowska et al. [13] examined the degradation of polyurethanes in the Baltic Sea water and in seawater with NaN_3 for periods longer than 12 months. They characterised changes in the weight, tensile strength, and morphology of the polyurethane specimens after different periods of immersion in both environments and demonstrated that the degree of polymer degradation in seawater depended on the degree of cross-linking. The polyurethane-metal interface was studied by Possart [19], who showed that the bulk properties of the adhesive dominate its behaviour, when the polyurethane film exceeds a certain width but that the polymer chemistry is modified

polymer is small, Fick's second law can be applied [25, 26] to calculate the diffusion coefficient of water in the resin:

$$\frac{dC}{dt} = D\frac{d^2C}{dx^2},$$

(1)

where C is the concentration of seawater absorbed (%); t is the immersion time (s); D is the diffusion coefficient of water in the adhesive (m^2s^{-1}); x is the depth of water penetration (m).

Absorbed fluid concentration C(t) is usually expressed as the relative difference between the wet and dry masses of the specimen. That is,

$$C(t) = \frac{M(t) - M_0}{M_0},$$

(2)

where M(t) is the mass of the specimen which has been immersed in water for a time t and M_0 is the mass of dry specimen.

The Fick diffusion curve is the curve of concentration versus time. The apparent diffusion coefficient D_A [25] can be calculated by applying it to the linear zone of Fick's diffusion curve, as shown in the following equation:

$$D_A = \frac{\pi}{16}\left(\frac{h(C(t_2) - C(t_1))}{C_s(\sqrt{t_2} - \sqrt{t_1})}\right)^2,$$

(3)

where C_s is the concentration of equilibrium or chemical saturation; $C(t_1)$ amount of fluid absorbed until the time t_1; $C(t_2)$ amount of fluid absorbed until time t_2; h thickness of the specimen in m (1.5 mm in our case).

The specimens were immersed in the Santander Bay (Spain) seawater, which had a pH of 8.2 at room temperature, for durations of 1, 2, 3, 8, 16, 32, 64, or 180 days. The amount of water absorbed was determined through gravimetric analysis. The sequence of the degradation processes that occur upon immersion in seawater is shown in Figure 1.

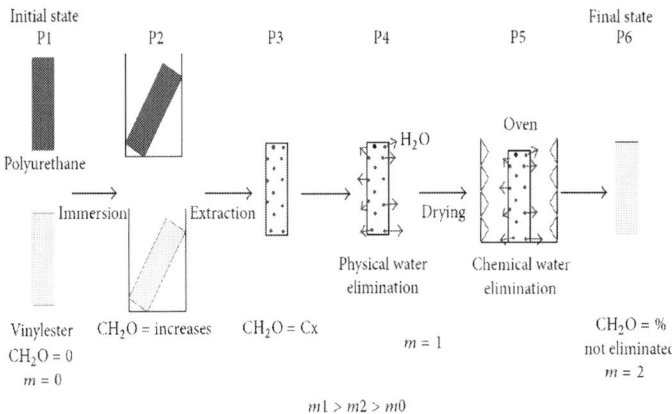

Figure 1: Degradation process.

The water absorbed by the polymer has been partially eliminated by vacuum drying the specimens. Drying is carried out by heating the specimens at a temperature of 40°C in a container where the vacuum has been done (0.053 Atm). The water that is chemically linked to the polar groups in the polymer will be retained after the vacuum drying and is responsible for the long-term degradation of the adhesive mechanical properties. The water that is not chemically linked temporarily modifies the polymer's properties, but these properties recover after its elimination during vacuum drying.

Analysis Methods

Several different techniques were used to follow the evolution of the polymer environmental degradation processes.

- For gravimetric analysis, an electronic scale, model SATRORIUS TE 214S, with a precision of 0.0001 g was used to measure the water uptake for determination of the adhesive diffusion coefficients.

- The depth to which seawater has diffused into the adhesive joint was determined by a scanning electron microscope (SEM), model JEOL JSM 5600. The specimens were metalized with gold by sputtering in an argon atmosphere, and an electron acceleration voltage of 20 kV was used for the imaging.

- ATR-FTIR spectroscopy, model BRÜCKER TENSOR 27, was used at room temperature in the 4000–525 cm^{-1} wavenumber range to study the chemical changes in the adhesive throughout the degradation process that is a result of extended exposure to seawater.

- A differential scanning calorimeter (DSC), model SDT Q600, was used to observe the change in the glass transition temperature (T$_g$) for the different degradation periods; the temperature range was from 0°C to 400°C under a nitrogen gas atmosphere, and the heating rate was 20°C/min.

- An IBERTEST IBTH3630 machine with a 200 N load cell and a movement rate of 2 mm/min was used to characterise the property losses that the adhesive experiences as an effect of the degradation.

RESULTS AND DISCUSSION

Diffusion Coefficients and Seawater Concentration Profiles

The diffusion coefficients of the adhesives were calculated using the gravimetric results obtained from the bulk adhesive specimens. In Figure 2, an initial period can be observed, in which the weight increases as a function of the time for which the specimen has been immersed because of the water absorption by the adhesive. These data were used to calculate the diffusion coefficient. From day 9 of immersion, the specimen weight begins to decrease due to the onset of the degradation phenomena in the polymer. We consider the concentration at which the weight maximum is reached to be the critical water level. In both adhesives and 9 days after the immersion, it was found that the weight of the specimen decreases. This occurs because the water attacks the polymer chains (which have operated since the beginning of immersion). These phenomena degrade irreversibly the adhesive because they produce bond breakage and loss of polymer fragments that pass to seawater.

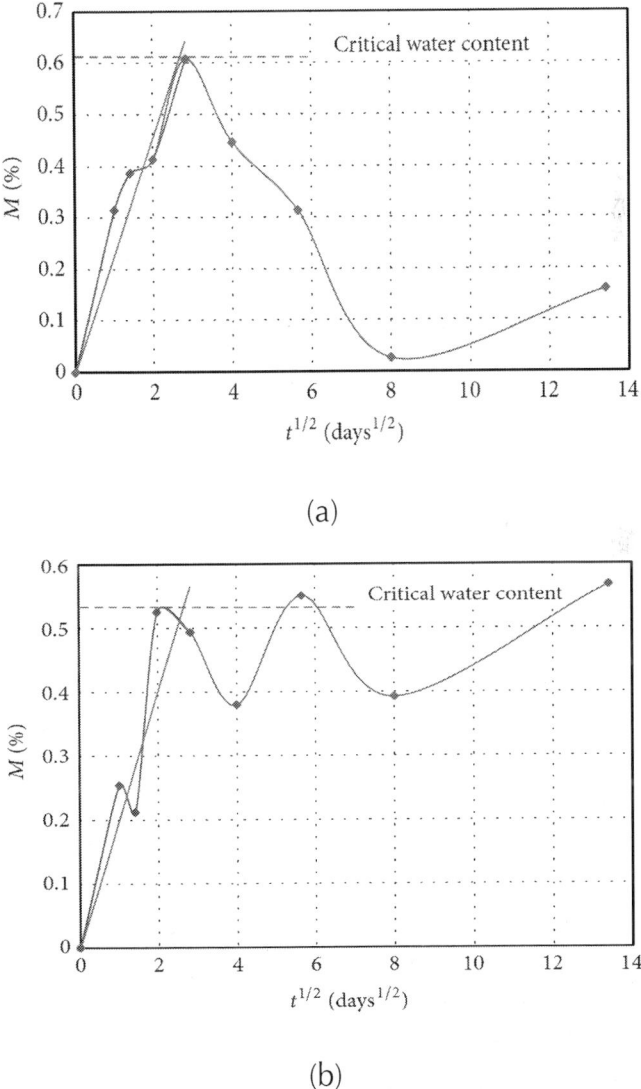

(a)

(b)

Figure 2: Water absorption curves as a function of specimen immersion time: (a) polyurethane and (b) vinylester.

From these data, the water diffusion coefficients of both polymers can be obtained. The fact that the adhesive specimen has 6 faces through which water can enter the polymeric network must be taken into consideration, making it necessary to use a correction factor to Fick's law to take into account the diffusion through the sides. Thus, the following expression is applied:

$$D = D_o\left(1 + \frac{h}{l} + \frac{h}{b}\right)^{-2},$$

(4)

where D is the water diffusion coefficient in the adhesive (m²s⁻¹); D_0 is the diffusion coefficient without considering water entry through the sides (m²s⁻¹);l is the specimen length (m); b is the specimen width (m); is the specimen thickness (m).

The measured diffusion coefficients for the adhesives are 6.39×10^{-13} and 5.14×10^{-13} m²s⁻¹ for polyurethane and vinylester, respectively. Once the diffusion coefficients are known, the concentration profiles of water in an adhesive joint with a given thickness can be obtained as a function of the immersion time. Thus, we can determine how long the water takes to diffuse into the centre of the adhesive joint and what amount of time is necessary for the water concentration in the central zone to exceed a critical value above which irreversible chemical changes occur, which result ultimately in the permanent loss of the mechanical properties of the joint. Bowditch [17] examined how water can affect the physical and mechanical properties of the adhesive and also the nature of the interface between it and the substrate. Bowditch considers the existence of a residual force at the joints (where the adhesive can be separated from the substrate by the action of water). He described the mechanisms that influence the durability of the polymeric adhesives in the presence of seawater. Datla et al. [28] examined the effect of mass adhesive samples subjected to different levels of humidity and 50°C, and they concluded that the saturation level is increased with the relative humidity. They proposed a model of absorption in two stages. This model gave excellent results and was easily modeled with FEM. The loss of mechanical properties in the adhesive joint was investigated with

Mixed Cantilever Beam (MCB) specimens in composite mode. This specimen configuration, proposed by Högberg and Stigh [29, 30], has advantages over other configurations commonly used in the study of the environmental degradation of adhesive joints.

The water concentration profiles as a function of immersion time were calculated using Fick's law and incorporated the geometry of the MCB specimens. Thus, the time needed for the centre of the specimen to reach the critical humidity level, the level required to cause irreversible damage to the polymer properties, can be determined. Times of 139 and 94 days for polyurethane and vinylester, respectively, were determined (Figure 3).

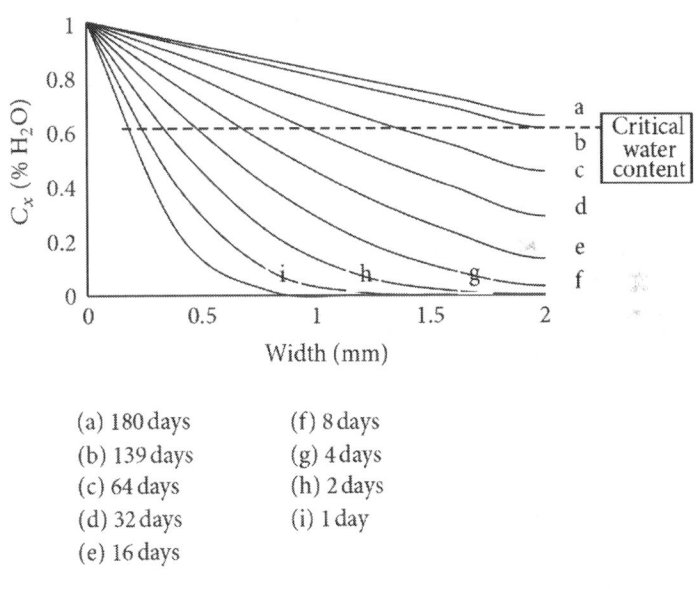

(a) 180 days	(f) 8 days
(b) 139 days	(g) 4 days
(c) 64 days	(h) 2 days
(d) 32 days	(i) 1 day
(e) 16 days	

(a)

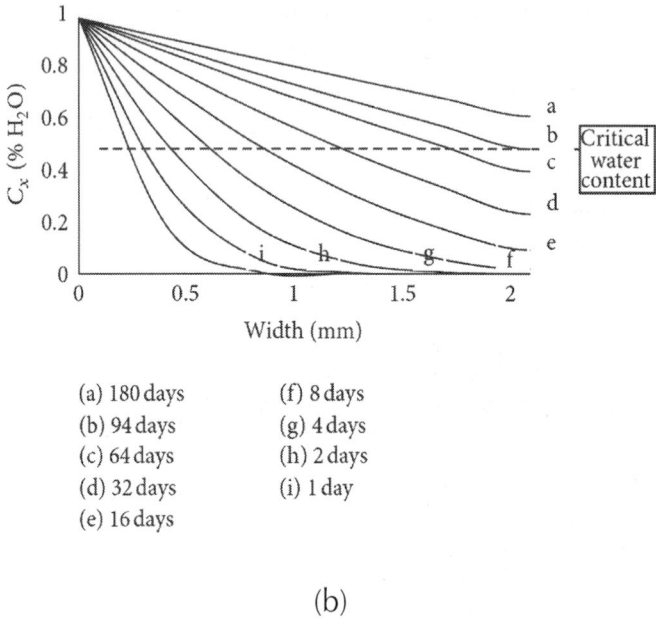

(a) 180 days   (f) 8 days
(b) 94 days   (g) 4 days
(c) 64 days   (h) 2 days
(d) 32 days   (i) 1 day
(e) 16 days

(b)

Figure 3: Concentration profiles of seawater in MCB test specimens as a function of immersion time: (a) polyurethane and (b) vinylester.

Scanning Electron Microscopy

The depth to which seawater has diffused into the adhesive joint was determined by SEM. It must be taken into account that water absorption occurs through all of the borders of the adhesive joint and diffuses throughout the immersion period. Figure 4 shows the penetration of water as a function of the immersion time. Due to the chemical action of the absorbed water, some clearly delineated zones can be observed, which mark the limit to which absorbed water has irreversibly changed the polymer structure during the time considered. The darker zones are regions where water has not attacked by hydrolysis the polymer chains; that is, the concentration of water has been at all times below the critical concentration (which causes irreversible changes in the polymer). The water absorbed by the adhesive produces microstructural changes in the polymer due to the hydrolysis of certain macromolecular functional groups.

The functional groups responsible for these chemical changes were identified by FTIR and will be presented later in this paper. These morphological changes are translated into losses of mechanical properties that have been investigated by creep tests on specimens Mixed Cantilever Beam (MCB) at different temperatures. After 180 days of immersion of the specimens, it was concluded that the -integral value (depending on the strain) implies a loss of stiffness of 51% and a decrease in the failure load of 59% for the adhesive tested.

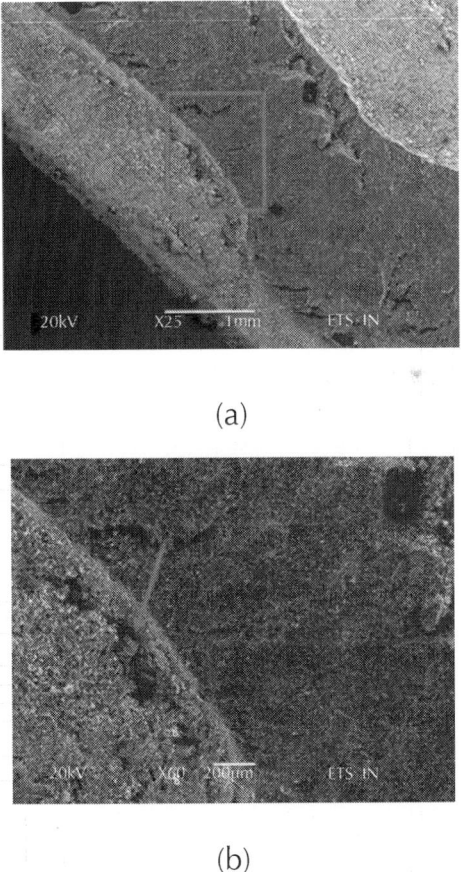

(a)

(b)

Figure 4: SEM image of the surface of polyurethane adhesive MCB specimens after fracture. Following a 32-day seawater immersion period, (a) the limit of the water diffusion zone and (b) a detailed image of the adhesive's degradation produced by the action of water.

Upon more careful observation of the adhesive surface (Figure 5), three well-differentiated zones can be identified, which are related to the water concentration reached and the chemical changes caused in the polymer.

- Zone 1 is where the absorbed water has not exceeded the critical concentration of water, and therefore the initial properties of the adhesive remain unchanged.

- Zone 2 is the region to which seawater has penetrated and concentrations were above the critical concentration, chemically linked to the polymer, and the polymer properties are irreversibly degraded.

- Zone 3 is the region in which salt deposits occur because the solubility product of dissolved salts in seawater has been reached locally. These deposits constitute a physical barrier to the ingress of water.

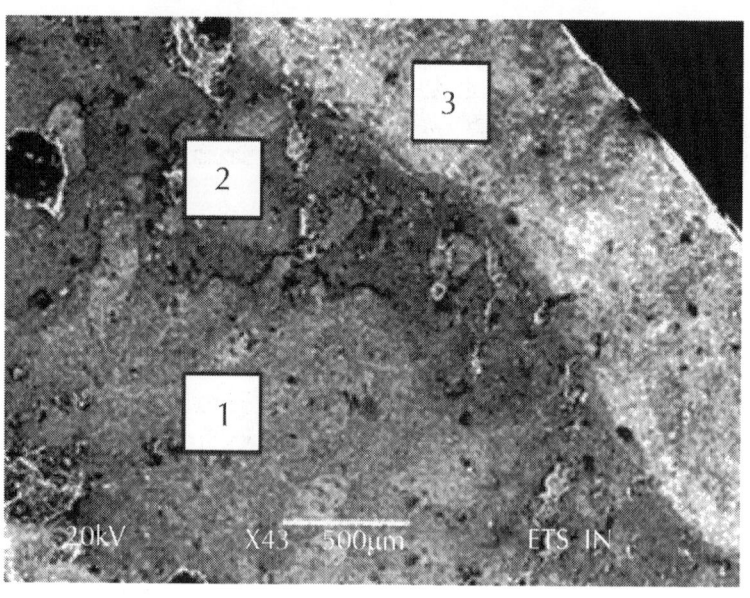

(a)

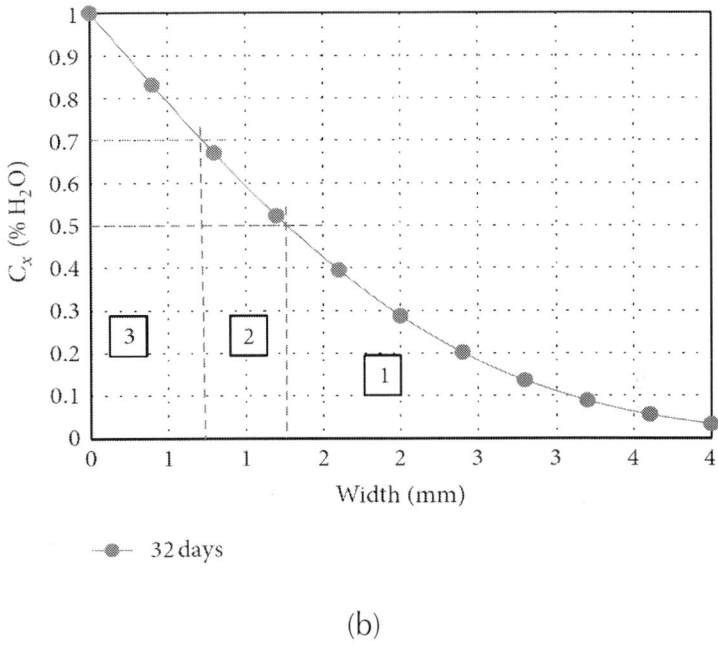

(b)

Figure 5: (a) SEM image of the surface of the polyurethane adhesive MCB specimens after fracture following a 32-day immersion period in seawater. (b) Different zones as a function of water concentration.

Polyurethane can absorb water to approximately 0.6% before irreversible degradation takes place. In Figure5(a), this limit corresponds to Zone 3. Upon closer SEM inspection, saline residues are visible on the polymeric network, which block the slide of macromolecules until the material cannot be deformed and finally fails under the loads placed on it (Figure 6). Tests have been performed for the diffusion of distilled and demineralized water in the polymer, and subsequently we have used natural seawater. In this case, the diffusion coefficients are slightly smaller. Since the water molecule is of the same size in both cases and the same average size of the polymer network and considering that has been cured in standard conditions, we propose the following hypothesis: decreases in the diffusion coefficients can be due to the action of saline sediments present in the polymer network. This working hypothesis is based on the discrepancy of the diffusion coefficients.

Indeed, it may cause precipitation of the salt compounds during the drying period in vacuum but, during the diffusion process can produce a precipitation localized because of the difficulty of homogenization of the salt concentration at each sampling point, as diffusive flow imposes a gradient in the concentration of water: the water molecules continue to advance so that the salt concentration increases in the zones that are being left behind until it exceeds the solubility product and the precipitated appears. This precipitated acts as a physical barrier to the entry of more water in the polymer.

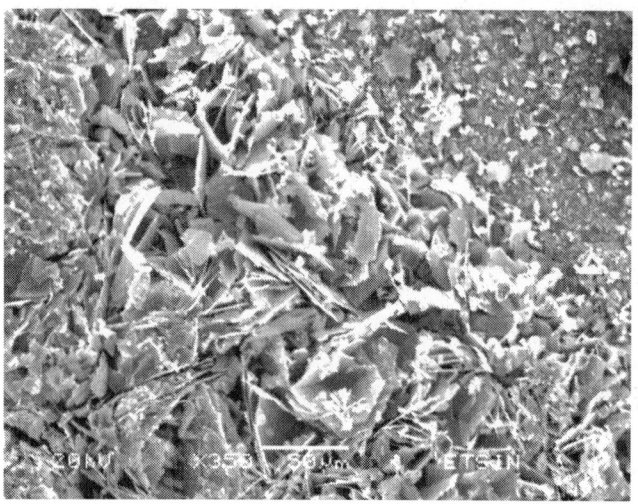

Figure 6: SEM image of the saline deposits on the polyurethane adhesive.

The fracture of the vinylester zone 3 material is shown in Figure 7, and it is different from that observed in polyurethane. After 32 days of water absorption, the water retained swells the macromolecular network, which causes tears in the material, together with a selective attack on some functional groups. This damage is permanent and is linked to reach a critical water content and to generate sufficiently high tension forces in the polymer to produce a fracture. The material experiences an irreversible degradation of its mechanical properties, as shown by the mechanical tests performed, as will be discussed later.

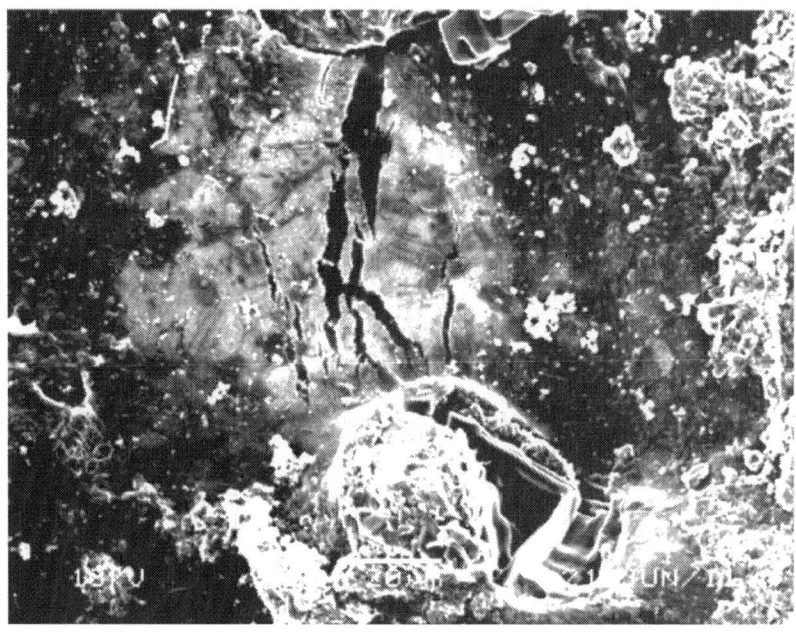

Figure 7: SEM image of vinylester resin with details of the blistering caused by a 32-day immersion in seawater.

The quantification of these phenomena would require mechanical tests in different humidity environments at different temperatures and with several levels of tension applied over the adhesive joint.

Infrared Spectroscopy

FTIR spectroscopy was performed to study the chemical changes that have been produced in the adhesive throughout the degradation process.

Figure 8(a) shows the polyurethane adhesive FTIR spectrum. In order to know the initial state of the adhesive, tests have been made with the adhesive intact, that is, without immersion in water.

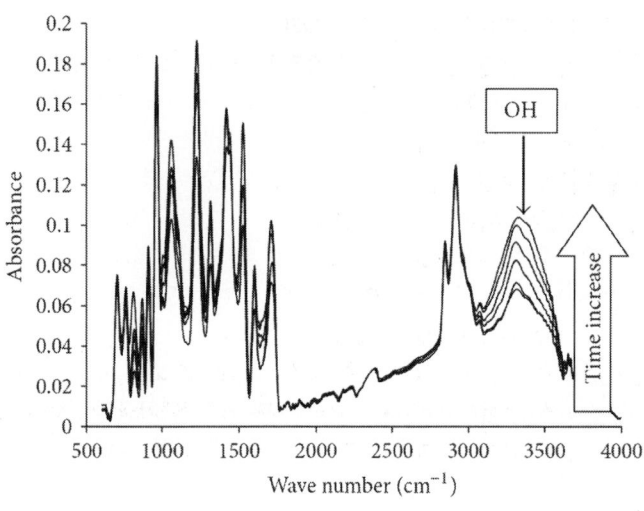

(a)

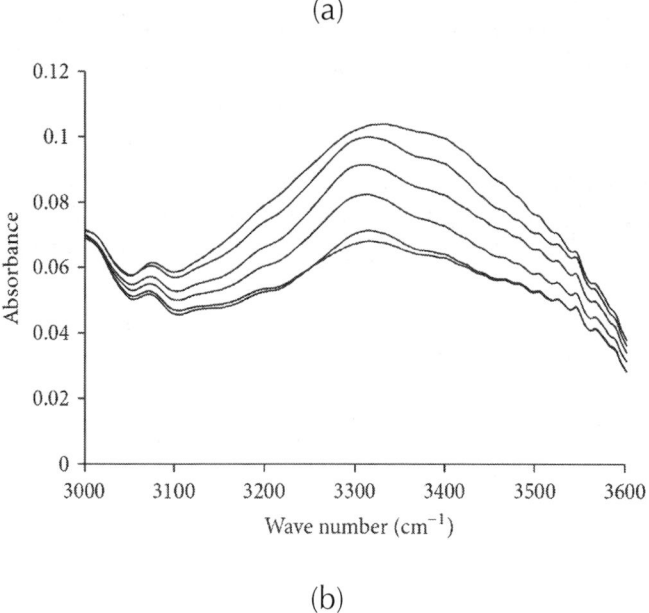

(b)

Figure 8: IR spectrogram: (a) polyurethane and (b) details of band 3400 cm⁻¹.

The bands at 2934 and 2862cm⁻¹ correspond to the C–H stretching in the polyurethane, and the bands due to N–H are found at 3334 and 1530cm^{-1}. Other absorption bands are N–H plane (1168 cm^{-1}) and N–H out-of-plane deformations (734 cm^{-1}). The band at 1724 cm^{-1} corresponds to the C = O stretching of urethane, and the bands at 1261 and 970 cm^{-1} are due to C–N stretching and C–O–C in urethane, respectively. Furthermore, the bands at 1219 and 1073 cm^{-1} correspond to antisymmetrical and symmetrical N–C=O stretching, respectively [31, 32]. The absorption band located within the 3400 to 3500 cm^{-1} range corresponds to the O–H linked by hydrogen bridges or to different groups linked to molecular water through hydrogen bonds. It is observed that, with an increase in the immersion time, the N–H band stretching changes to higher frequency values due to the water absorbance that is occurring [27].

Figure 9(a) shows the FTIR spectrum of vinylester. The band at 3416 cm^{-1} corresponds to the O–H stretching vibrations in the vinylester. The band at 3036 cm^{-1} corresponds to the C–H stretch of the benzene ring, and the bands due to C=C are found at 1635 and 940 cm^{-1}. Another absorption band due to benzene is at 1889 cm^{-1}, which corresponds to aromatic ring vibration, while the aromatic ring stretch is at 1581 cm^{-1}. The band at 1712 cm^{-1} corresponds to the C=O stretching, and the bands at 1233 and 1160 cm^{-1} are due to C–O–C and C–CO–O stretching, respectively. Furthermore, the bands at 1038 and 767 cm^{-1} correspond to aromatic C–H bending and aromatic ring stretch, respectively [33]. As in the case of the polyurethane adhesive, an absorption band located within the 3400 to 3500 cm^{-1} range is observed after immersion in seawater, which corresponds to O–H groups linked by hydrogen bridges or to other groups linked to molecular water through hydrogen bonds. An increase in frequency is also observed in the 3400 to 3500 cm^{-1} band due to the formation of hydrogen bonds [27].

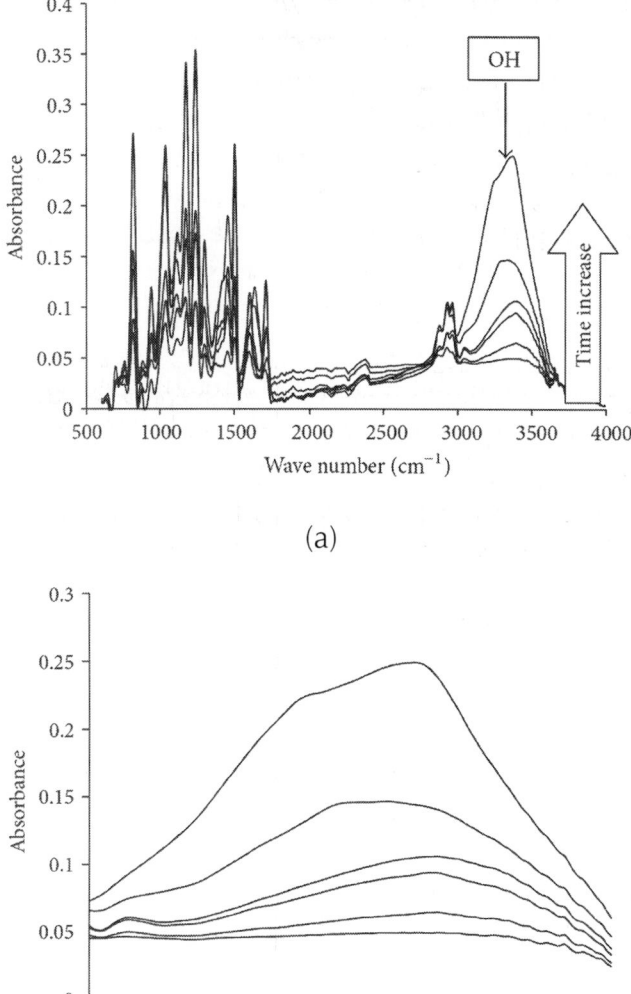

Figure 9: IR spectrogram: (a) vinylester and (b) details of band 3400 cm⁻¹.

The analysis of the polyurethane adhesive samples throughout the degradation period did not show significant changes from the reference spectrum, except for an increase in the 3400 cm⁻¹ band as the immersion time increases, which corresponds to the O–H groups linked by hydrogen bridges due to the water absorption that occurs. However, the vinylester resin spectral analysis shows many changes throughout the immersion period, which indicates the many degradation processes that occur; this result was expected because vinylester resin is a crosslinked polymer. As in the polyurethane adhesive, an increase in the 3400 cm^{-1} band occurs due to the hydrolysis that is experienced, which becomes more pronounced as the immersion time increases.

A broadening of the peak corresponding to the 3400 cm^{-1} band for each of the polymers is shown in Figures 8(b) and 9(b). This band is caused by the entry of water. In the case of polyurethane joins NH groups and is chemically bound. It is responsible for the degradation of the mechanical properties. Figure 10 shows the effect of water on the chemical structure of the polyurethane.

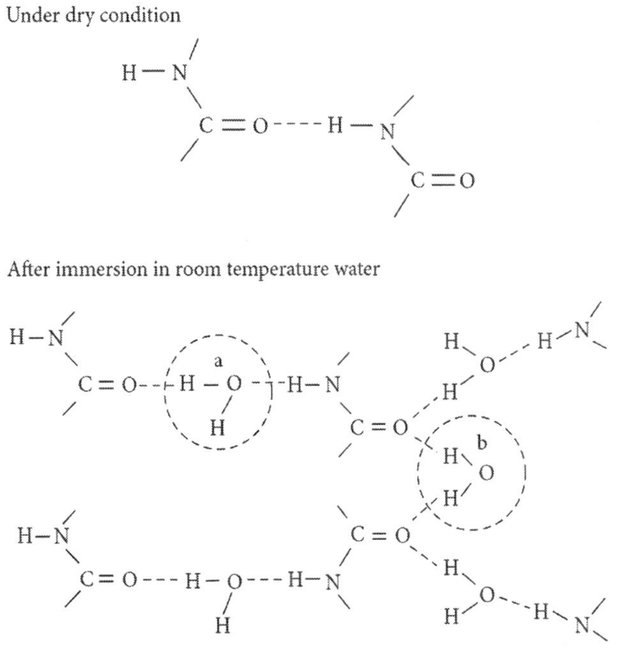

Figure 10: Effects of water on the hydrogen bonding in polyurethane SMP [27].

Differential Scanning Calorimetry (DSC)

Through differential scanning calorimetry, the change in the glass transition temperature (T_g) was observed in the different degradation periods. Figure 11 shows the DSC traces for a pristine vinylester resin sample and a sample after 9 months of immersion. The glass transition temperature is determined automatically by the computer control software, which determines the midpoint between the two marks indicated on the curve. This midpoint corresponds to T_g for each sample analyzed.

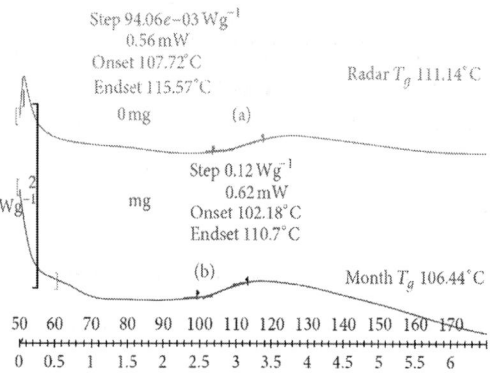

Figure 11: DSC of a 5 mg vinylester sample: (a) pristine sample and (b) after 9 months of degradation.

A reduction in the glass transition temperature value is observed after 9 months of degradation in seawater (T_g of a pristine sample = 111.41°C and T_g after 9 months of degradation = 106.44°C). It is also observed that the vinylester resin without degradation is stable until approximately 180°C; after 9 months of degradation, however, it does not stabilise, and instead it drops continuously after the glass transition, indicating that additional degradation is taking place during heating. Significant changes in the polyurethane glass transition temperatures are not observed after a long immersion time due to the high pigmentation of polymer, which makes T_g acquisition difficult.

Tensile Test

The results of the tensile test are shown in Figure 12. A reduction in the mechanical properties was observed after a long immersion time in natural seawater for both resins. In the vinylester resin, the tensile strength decreases from 120.95 N before immersion to 107 N after 9 months of immersion in seawater. For polyurethane, the decrease in tensile strength is from 24.81 N before immersion to 19.7 N after 9 months of immersion. Degradation of the resin molecular structure clearly occurs, causing irreversible damage and a reduction of the mechanical properties. A tension test alone is not enough to fully characterise the loss in mechanical strength experienced by the adhesive as a result of the degradation caused by exposure to seawater. The polymers have some viscoelastic behaviour, and the tensile test results do not reflect the loss in resistance capacity under constant loads applied over long-time periods. The performance of creep assays.is precise.

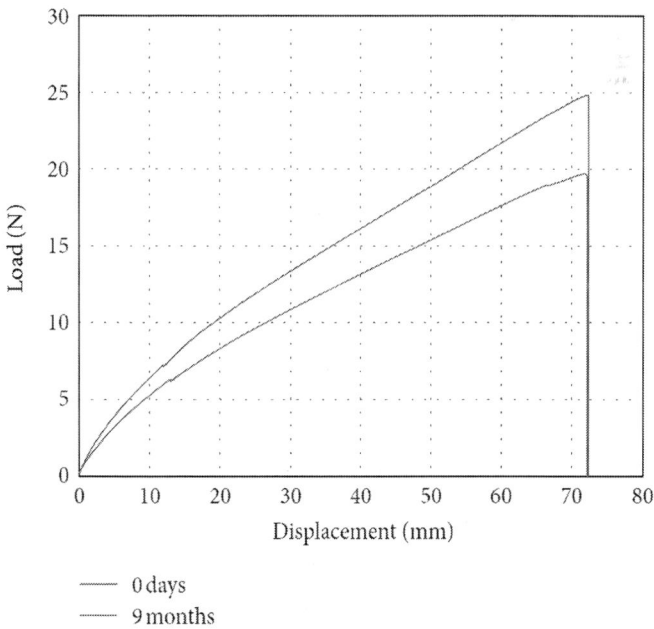

(a)

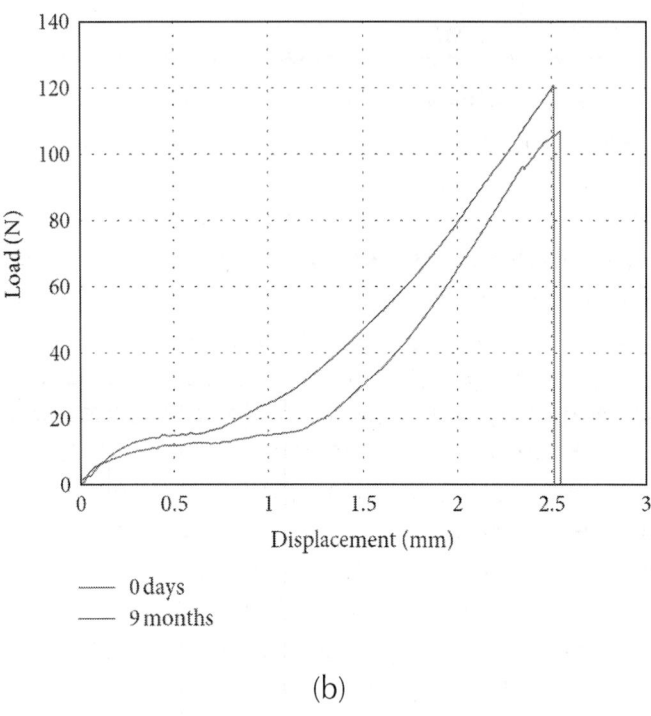

(b)

Figure 12: Tensile test load-displacement plots for (a) polyurethane and (b) vinylester.

CONCLUSIONS

- The water concentration profiles in adhesive joints were determined for both polyurethane and vinylester adhesives using diffusion coefficients calculated via gravimetric methods. The diffusion coefficient is 6.39×10^{-13} m^{2}s^{-1} for polyurethane and 5.14×10^{-13} m^{2}s^{-1} for the vinylester. According to these values, the time required for the water concentration in the central zone to exceed a critical value, above which irreversible damage in the polymer occurs, can be determined.

- The microstructural changes in the polymers due to the hydrolytic action of water were observed by scanning electron microscopy (SEM), and three distinct zones corresponding to different water concentrations reached and the chemical modifications caused in the polymer were identified.

- In the infrared absorption (FTIR) spectra of both polymers, increases in the $3400\,cm^{-1}$ band corresponding to bonded OH groups were observed; this increase was attributed to water absorption in the polymeric network with increasing immersion time.

- In the vinylester resin FTIR spectrum, large changes in several other spectral bands occur throughout the degradation period, which indicates a relatively greater alteration of the polymer molecular structure. Hydrolytic attack to different functional groups of the molecule appears to result in faster degradation and a greater final water concentration inside the polymer network than with polyurethane.

- A reduction of the vinylester resin glass transition temperature from a pristine sample to one subjected to nine months of degradation was observed using DSC. In addition, a continuous decrease in the curve after the glass transition was observed, indicative of further degradation. An appreciable change in the glass transition temperature of polyurethane was not observed.

- The creep tests (that were carried out with MCB specimens) have allowed us to calculate the loss in stiffness and strength of the adhesive joints in function of immersion time in natural seawater. In this way, we calculated the evolution of the j-integral in function of the strain of the adhesive during the testing time. In the case of polyurethane, we calculated a decrease of 51% in stiffness and a 59% in strength.

ACKNOWLEDGMENTS

The authors would like to thank the Madrid Polytechnic University (Universidad Politécnica de Madrid) for the project N°123 and the TEKHIMAT company for the funding received for carrying out

17. M. R. Bowditch, "The durability of adhesive joints in the presence of water," International Journal of Adhesion and Adhesives, vol. 16, no. 2, pp. 73–79, 1996.

18. J. Burns, P. S. Dubbelday, and R. Y. Ting, "Dynamic bulk modulus of various elastomers," Journal of Polymer Science B, vol. 28, no. 7, pp. 1187–1205, 1990.

19. W. Possart, "Chemical structura formation and morphology in ultrathin polyurethane films on metals," in Adhesion, Current Research and Applications, vol. 6, pp. 71–88, WILEY-VCH, Weinheim, Germany, 2005.

20. P. Davies, F. Mazéas, and P. Casari, "Sea water aging of glass reinforced composites: shear behaviour and damage modelling," Journal of Composite Materials, vol. 35, no. 15, pp. 1343–1372, 2001

21. J. Raghavan and M. Meshii, "Creep rupture of polymer composites," Composites Science and Technology, vol. 57, no. 4, pp. 375–388, 1997.

22. E. Woo, "Moisture-temperature equivalency in creep analysis of a heterogeneous-matrix carbon fibre/epoxy composite," Composites, vol. 25, no. 6, pp. 425–430, 1994

23. D. J. O'Brien, P. T. Mather, and S. R. White, "Viscoelastic properties of an epoxy resin during cure,"Journal of Composite Materials, vol. 35, no. 10, pp. 883–904, 2001.

24. J. C. Suárez, C. Alia, M. V. Biezma, P. Pinilla, and J. M. Arenas, "Long term ageing of adhesives for hybrid materials in seawater," in Proceedings of the 9th European Adhesion Conference (EURADH '12), p. 44, Friedrichshafen, Germany, 2012.

25. D. A. Dillard, Advances in Structural Adhesive Bonding, Woodhead, 2010.

26. J. Crank and G. S. Park, Diffusion in Polymers, Academic Press, 1968.

27. B. Yang, W. M. Huang, C. Li, and L. Li, "Effects of moisture on the thermomechanical properties of a polyurethane shape

memory polymer," Polymer, vol. 47, no. 4, pp. 1348–1356, 2006.

28. N. V. Datla, M. Papini, J. Ulicny, B. Carlson, and J. K. Spelt, "The effects of test temperature and humidity on the mixed-mode fatigue behavior of a toughened adhesive aluminum joint," Engineering Fracture Mechanics, vol. 78, no. 6, pp. 1125–1139, 2011.

29. J. Högberg and U. Stigh, "Specimen proposals for mixed mode testing of adhesive layer," Engineering Fracture Mechanics, vol. 73, no. 16, pp. 2541–2556, 2006

30. J. C. Suárez, P. Pinilla, F. López, M. A. Herreros, and M. V. Biezma, "Mixed mode double cantilever beams test specimen for characterization of structural adhesive joints," Anales De Mecánica De Fractura, vol. 27, p. 685, 2010.

31. R. Torregrosa, S. Álvarez, and J. M. Martín, "Migration of low molecular weight moiety at rubber-polyurethane interface: an ATR-IR spectroscopy study," International Journal of Adhesion and Adhesives, vol. 31, no. 6, pp. 389–397, 2011.

32. M. Bassyouni, S. A. Sherif, M. A. Sadek, and F. H. Ashour, "Synthesis and characterization of polyurethane—treated waste milled light bulbs composites," Composites B, vol. 43, no. 3, pp. 1439–1444, 2011.

33. N. Jost and J. Karger-Kocsis, "On the curing of a vinylester-urethane hybrid resin," Polymer, vol. 43, no. 4, pp. 1383–1389, 2001.

Structural Health Monitoring of Civil Infrastructure Using Optical Fiber Sensing Technology: A Comprehensive Review

X. W. Ye[1], Y. H. Su[2], and J. P. Han[2]

[1]Department of Civil Engineering, Zhejiang University, Hangzhou 310058, China

[2]Key Laboratory of Disaster Prevention and Mitigation in Civil Engineering of Gansu Province, Lanzhou University of Technology, Lanzhou 730050, China

ABSTRACT

In the last two decades, a significant number of innovative sensing systems based on optical fiber sensors have been exploited in the engineering community due to their inherent distinctive advantages such as small size, light weight, immunity to electromagnetic interference (EMI) and corrosion, and embedding capability. A lot of optical fiber sensor-based monitoring systems have been developed for continuous measurement and real-time assessment of diversified engineering structures such as bridges, buildings, tunnels, pipelines, wind turbines, railway infrastructure, and geotechnical structures. The purpose of this review article is devoted to presenting a summary of the basic principles of various optical fiber sensors, innovation in sensing and computational methodologies, development of novel optical fiber sensors, and the practical application status of the optical fiber sensing technology in structural health monitoring (SHM) of civil infrastructure.

INTRODUCTION

Structural health monitoring (SHM) has been a fast-developing domain in engineering disciplines especially in civil engineering field. The innovation in the SHM technologies as well as the development of the large-scale SHM systems has boomed within the engineering and academic communities over the last two decades [1–7]. The available practical experiences have proved that the progressive advancement of the sensing techniques will undoubtedly expedite the evolution of the SHM technology. In comparison with the traditional mechanical and electrical sensors, the optical fiber sensors possess some unique advantages such as small size, light weight, immunity to electromagnetic interference (EMI) and corrosion, and embedding capability [8–12], and therefore they have been employed in monitoring of engineering structures worldwide. This paper will provide a comprehensive review on structural monitoring of civil infrastructure by use of the optical fiber sensing technology. In the last two decades, a considerable number of investigations have been conducted in reviewing the progress of research and development of the optical fiber sensing technology as well as the applications of optical fiber sensors in the monitoring of various kinds of engineering structures [13–17]. Bhatia et al. [18] reported the progress in the performance and reliability of the optical fiber-based extrinsic Fabry-Perot interferometric (EFPI) strain sensor. Rao [19] gave a systematic review of progress on applications of FBG sensors in large composite and concrete structures, the electrical power industry, medicine, and chemical sensing. Leung [20] reviewed the applications of optical fiber sensors for monitoring of concrete structures. Measures et al. [21] overviewed the research on the development of structurally integrated optical fiber sensors for the smart structures. Merzbacher et al. [22] reviewed the strain monitoring of concrete structures by use of optical fiber sensors. López-Higuera et al. [23] summarized the main types of optical fiber techniques suitable for structural monitoring and introduced various optical fiber sensor-based engineering application scenarios. Ansari [24] provided a summary of basic principles pertaining to monitoring of civil engineering structures with optical fiber sensors. Majumder et al. [25] reviewed the recent research and development activities in structural monitoring using FBG sensors.

FUNDAMENTALS OF OPTICAL FIBER SENSORS

Generally, an optical fiber sensor system consists of a light transmitter, a receiver, an optical fiber, a modulator element, and a signal processing unit. As the core part of an optical fiber sensor, the optical fiber is usually made from silica glass or polymer material, which itself can act as a sensing element or carry the light from the source to the modulator element. When the strain or temperature variation of the structure occurs, the surface-mounted or embedded optical fiber sensor in the structure will expand or contract. In accordance with the change of the length of the optical fiber, the optical fiber sensor modulates the light and reflects back an optical signal to the analytical unit for deriving the concerned physical quantity of the structure [26]. Based on the sensing principle, the optical fiber sensors can be categorized into different types as illustrated in the following sections [27].

Fiber Bragg Grating (FBG) Sensors

Up to now, the FBG sensor has been widely used in the monitoring of civil engineering structures [28–32]. It can be regarded as a type of optical fiber sensor with varied refractive indices in the core. According to Bragg's law, a beam of white light is written in the FBG sensor, and when the light from the broadband source passes through the grating at a particular wavelength, the Bragg wavelength is reflected which is related to the grating period, as illustrated in Figure 1. The Bragg wavelength λ_B can be expressed by

$$\lambda_B = 2n_{eff}\Lambda,$$

(1)

Where n_{eff} is the effective index of refraction and Λ is the grating period. The wavelength shift changes linearly with both strain and temperature. When the grating part is subjected to external disturbance, the period of the grating will be changed and the Bragg wavelength is varied accordingly. The variation of the Bragg wavelength can be obtained by

$$\Delta\lambda_B = \lambda_B \left\{ (\alpha + \xi)\, \Delta T + (1 - p_e)\, \Delta\varepsilon \right\},$$

(2)

Where $\Delta\varepsilon$ is the strain variation, ΔT is the temperature change, α is the coefficient of the thermal expansion, ξ is the thermooptic coefficient, and P_e is the strain-optic coefficient.

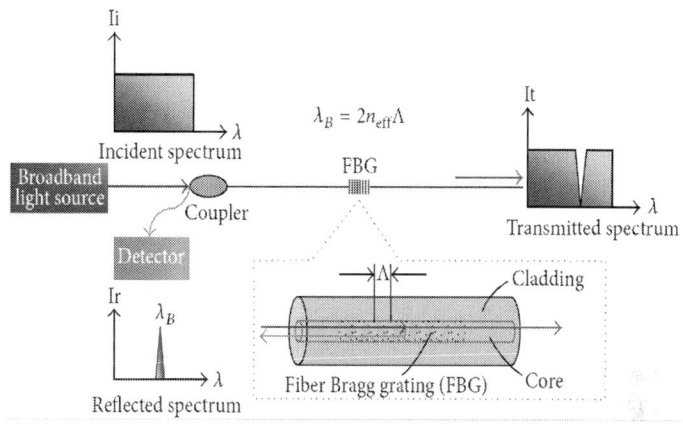

Figure 1: Measurement principal of FBG sensor.

Extrinsic Fabry-Perot Interferometric (EFPI) Sensors

For an EFPI sensor, the optical fiber acts as the input or output path; the light from the source passes through the optical fiber to the sensing part and then to the demodulation system [33–38]. A typical EFPI sensor consists of the input/output fibers and the reflective fibers as well as a hollow-core tube for creating an air cavity, namely, the Fabry-Perot cavity. An adhesive is employed to bond the two components. As illustrated in Figure 2, the Fabry-Perot cavity is formed between an input single-mode fiber and a reflecting single-mode or multimode fiber, and two fibers are aligned inside a hollow-core tube. At both ends of the cavity, there are reflections on the uncoated ends of the fibers. R_1 is referred to as the reference reflection which depends on the applied perturbation such as strain and temperature. R_2 is the sensing reflection and depends on the length of the cavity, L. A sinusoidal output signal will be generated when R_1 interferes with R_2. Because the length of the cavity can be modulated by the applied perturbation, the EFPI sensor can be used to measure the applied perturbation according to the output signal. For the strain measurement, it can be expressed by

$$\varepsilon = \frac{\Delta l\,(\text{air\ gap})}{L},$$

(3)

Where Δl is the variation in the cavity.

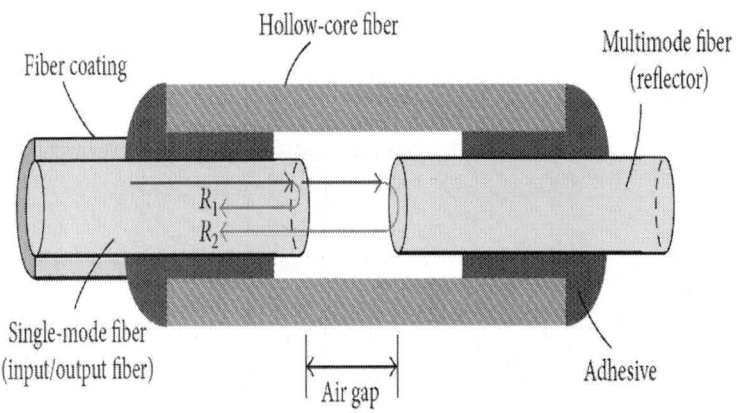

Figure 2: Measurement principal of EFPI sensor.

Optical Time-Domain Reflectometry (OTDR) Sensors

An optical time-domain reflectometry (OTDR) based sensor is capable of distributedly sensing along the length of an optical fiber with a specific refractive index [39, 40]. When a light pulse at a particular wavelength propagates along the optical fiber, the sensor can locate the position of the interaction according to the propagation time, as illustrated in Figure 3. The location of the variation of the measurand may be determined by the OTDR sensor. The OTDR-based distributed sensor is possible to be used to measure the change in the properties of the light along the entire optical fiber by measuring the time of flight of the returned pulses. The Brillouin optical time domain reflectometer (BOTDR) sensor is one of the distributed optical fiber sensors and is based on the Brillouin scattering. Due to the advantage of being capable of measuring continuous strain and temperature over a long distance, the BOTDR sensor has been widely applied in distributed monitoring of large-scale civil structures.

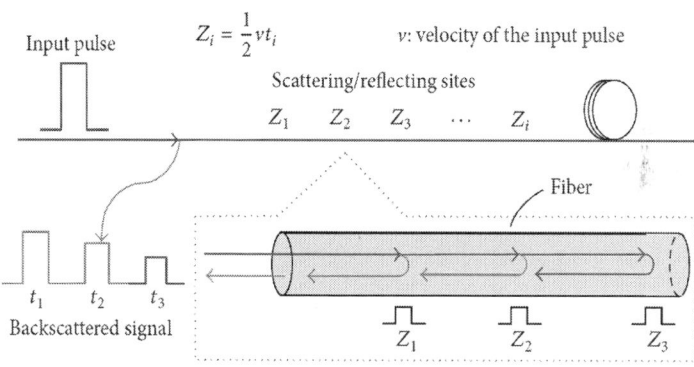

Figure 3: Measurement principal of OTDR sensor.

INNOVATION IN METHODOLOGIES AND SENSORS

Improvement of Methods

Li et al. [41] proposed a theoretical model for describing the strain transfer relationship between the fiber core of the FBG sensor and the host material. Yun et al. [42] developed a new method based on the simulated annealing evolutionary algorithm for demodulation of the strain profile along an FBG distributed strain sensor. Imai and Feng [43] proposed a stress-transferring model incorporating drastic softening behavior for the surrounding components to investigate the stress transfer from a host structure to a sensing fiber. Zhang et al. [44] proposed a model reconstruction soft computing recognition algorithm based on genetic algorithm, support vector regression to achieve the reliability of the FBG-based sensor network. Gill et al. [45] presented a genetic algorithm for the inversion of Bragg grating sensor spectral data to determine the strain distribution along the grating. Prabhugoud and Peters [46] developed an integrated formulation for the calculation of the spectral response of an FBG sensor embedded in a host material system as a function of the loading applied to the host structure.

Liu et al. [47] proposed an adaptive filtering algorithm for the noise reduction and the detectability of seismic signals measured by an FBG measuring system. Ma et al. [48] presented a fast interrogation method for the dynamic and/or static strain gauge by use of a reflection spectrum from two superimposed FBGs. Jang et al. [49] developed a real-time impact localization algorithm for various composite structures using the impact-induced acoustic signals acquired by multiplexed FBG sensors. Schizas et al. [50] proposed a method for nonhomogeneous strain monitoring of composite structures with embedded wavelength multiplexed FBG sensors. Feng et al. [51] proposed a stationary wavelet transform method for signal processing of the distributed strain data from the BOTDR-based optical fiber sensors. Peters et al. [52] investigated the method of measurement of nonuniform strain field near the stress concentration by use of the embedded FBG sensor.

Development of Sensors

Lee et al. [53] developed an optical fiber accelerometer composed of a reflective grating panel and two optical fibers as transceivers which was capable of monitoring the low-frequency acceleration of large-scale civil engineering structures. Wang and Huang [54] developed an optical fiber corrosion sensor based on the principle of light reflection consisting of an optical fiber reflection sensor and a tube/film subassembly formed by welding a sacrificial metallic film to a steel tube. Rodriguez-Cobo et al. [55] designed an FBG-based smart structure embedded into composite laminates for simultaneous measurement of temperature and strain and conducted experimental investigations for performance validation. Pirozzi [56] developed a multipoint force sensor based on crossed optical fibers. Kim et al. [57] developed a gold-deposited extrinsic Fabry-Perot interferometer for dynamic strain measurement.

Cumunel et al. [58] investigated the capacity of continuously attached long-gauge optical fiber sensors for dynamic evaluation of structures. Gangopadhyay et al. [59] addressed different design and experimental packaging procedures of indigenously developed FBG sensors for strain measurement. Yuan et al. [60] presented an optical fiber two-dimensional sensing system for measuring the strain inside the concrete structure based on white-light Michelson interferometric optical fiber sensing technique. Liu et al. [61] designed a long-period fiber grating sensor for detecting the state of rebar corrosion in concrete. Yashiro et al. [62] proposed an embedded chirped FBG sensor for damage detection in the holed carbon fiber reinforced polymer (CFRP) laminate. Zhou et al. [63] designed an extrinsic Fabry-Perot interferometric strain sensor for damage evaluation of smart composite beams. Triollet et al. [64] proposed a superimposed FBG device to measure, localize, and discriminate strain and temperature effects simultaneously for structural monitoring.

Hong et al. [65] developed a distributed long-gauge FBG macrostrain sensor for condition assessment of reinforced concrete beams [66]. Quirion and Ballivy [67] validated the robustness of the Fabry-Perot optical fiber sensor in strain monitoring of the concrete structure. Davis et al. [68] developed an integrated FBG-based sensing system for broad area damage detection of composite structures. Villalba and Casas [69] evaluated the usefulness and effectiveness of the optical backscatter reflectometer sensor in damage detection of concrete structures. Torres et al. [70] presented a new FBG strain sensor with an unsymmetrical packaging configuration designed to be fixed to the surface of the monitored structure. Schröder et al. [71] developed a low-cost optical fiber sensing system for continuous on-site monitoring of contact forces in current collectors. Kim [72] developed a smart monitoring system for stress monitoring and damage detection of offshore

sensing system using a compact FBG-based monitoring system incorporating a scanning Fabry-Perot filter, and the performance of the system was validated through laboratory experiments and field tests in a real bridge. Kister et al. [88, 89] conducted the research on structural monitoring of a composite road bridge by use of FBG sensors, and the performance of the adhesives and the protection system of the sensors were evaluated through field pullout tests. Mokhtar et al. [90] created an innovative FBG-based sensor system for accurate strain measurement with full temperature compensation towards condition monitoring and assessment of arch bridges. Surre et al. [91] developed an optical fiber sensor system for long-term strain monitoring and condition assessment of a redundant 50-year-old concrete footbridge.

In Hong Kong, the high-speed demultiplexing/interrogation system for FBG sensor arrays and FBG sensor package units were deployed for long-term monitoring on the Tsing Ma Bridge which is the world's longest suspension bridge carrying both highway and railway traffic [92]. Yau et al. [93] proposed a simple, inexpensive, and practical method for measurement of the vertical displacement of bridges by use of FBG sensors. In Chinese mainland, Zhao et al. [94] integrated the distributed Brillouin optical time domain analysis (BOTDA) technology and the FBG sensing technology for strain monitoring of bridges. In Korea, Chung et al. [95] conducted the experimental study on the applicability of long-gauge optical fiber sensors for the monitoring of the structural defection of the prestressed concrete bridges. Lin et al. [96] developed an FBG-based sensing system for online monitoring of highway bridges during construction to record the hydration effects, curing periods, prestressing responses, and removal of support frames.

Monitoring of Rehabilitated and Antique Bridges

Research efforts also have been devoted to measuring the structural behaviors of old bridges or deficient bridges during rehabilitation by use of the optical fiber sensing system [97]. Jiang et al. [98]

applied two types of optical fiber sensors embedded in FRP material to monitor the global and local behaviors of the strengthened bridge structures. Zhang et al. [99] introduced two types of optical fiber sensing technologies (FBG and BOTDR) for health monitoring of rehabilitated reinforced concrete girder bridges, and the static and dynamic loading tests were carried out with a simply supported reinforced concrete T-beam strengthened by externally posttensioned aramid fiber reinforced polymer (AFRP). Costa and Figueiras [100] presented the design of an advanced FBG-based monitoring system which was applied to a century steel arch bridge in Portugal.

Monitoring of Bridge Cables and Suspenders

He et al. [101] carried out an investigation on cable force monitoring by use of the local high-precision FBG sensor in combination with the distributed BOTDA sensing technique. Li et al. [102] developed a smart stay cable assembled with FBG-based strain and temperature sensors which were incorporated into a glass fiber reinforced polymer (GFRP) bar. The efficiency of the developed smart stay cable was proved by application to evaluate the fatigue accumulative damage of a stay cable bridge in China [103, 104].

Bridge Scour Monitoring

Zhou et al. [105] proposed an FBG sensing system for scour monitoring of foundations of bridge piers and abutments. This developed system introduced a uniform-strength FRP beam instrumented with two FBG sensors in two sides of the neutral axis, and the feasibility of the system was validated through laboratory tests. Lin et al. [106, 107] developed two types of FBG-based systems for real-time bridge scour monitoring, which were capable of measuring the process of scouring/deposition and the variation of the water level. The in situ FBG scour monitoring system was demonstrated to be robust and reliable for real-time scour-depth measurement and to be valid for indicating the depositional depth.

Xiong et al. [108] developed a bridge scour monitoring system by use of FBG sensors, and the experimental investigations verified that the recommended scour monitoring system was capable of measuring the water level, the scour depth, the entire process of scour development, and the deposition height due to the refilling process.

Buildings

The optical fiber sensing technology has been employed in safety condition monitoring of the high-rise structures during in-construction and in-service stages. Bastianini et al. [109] utilized the embedded optical fiber Brillouin sensors for strain monitoring and crack detection of a historical building. Antunes et al. [110] conducted dynamic monitoring of a reinforced concrete water reservoir and a slender metallic telecommunication tower by use of FBG-based biaxial accelerometers. Ni et al. [111] deployed massive FBG sensors for strain and temperature monitoring of the Canton Tower. Li et al. [112] performed an investigation on the feasibility of the FBG-based monitoring system instrumented in an 18-floor tall building during construction. The FBG sensors were used to monitor the strain and temperature of the building in three steps of construction, that is, before the concrete pouring, during the pouring and curing of concrete, and the construction of subsequent upper floors of the building.

Tunnels and Pipelines

Ye et al. [113] addressed two engineering paradigms on safety monitoring of tunnel construction by use of FBG sensors. Metje et al. [114] presented a new optical fiber sensing system for structural displacement monitoring which was successfully applied to measure the displacement of a tunnel lining. The novel system was based on a square fiberglass smart rod which was proved sensitive enough to measure the rotational movement of 0.5° and the lateral movement of 0.1 mm of the fixings. Li et al. [115] developed a metal groove encapsulating technique for the bare FBG sensor to measure the surface strain of the second lining of the tunnel. Li et al. [116] developed a differential FBG strain sensor for monitoring the stability of the tunnel during the backfilling and traffic-operating periods.

Glisic and Yao [117] proposed a method for real-time, automatic, or on-demand assessment of health conditions of buried pipelines after the earthquake based on distributed optical fiber sensors, the research of determination of sensor topologies, selection and development of sensors, development of installation and implementation procedures, and large-scale tests were conducted. Zhang et al. [118] experimentally investigated the prediction of locations and progression sequences of the pipe buckling with the aid of the broadening factor of the Brillouin spectrum width using high strength carbon-coated fibers and standard communication fibers.

Wind Turbines

A review of the current status and a discussion on research and implementation of FBGs and long-period gratings in wind turbine blade sensors can be found in [119]. Arsenault et al. [120] developed an FBG-based distributed strain sensor system for real-time monitoring of a wind turbine and conducted the validation tests under a laboratory scale under various loading conditions. Kim et al. [121] conducted experimental investigations on deflection estimation of wind turbine blades using embedded FBG sensors. Burgmeier et al. [122] developed and tested an FBG-based sensor system for remote measurement of strain that affects the power cable in an offshore wind turbine. Bang et al. [123] introduced an FBG-based arrayed sensor system for measurement of strain and bending deflection of wind turbine towers. Ge et al. [124] developed a specific intensity-modulated optical fiber accelerometer for vibration monitoring of wind turbine blades. Schroeder et al. [125] installed an FBG measurement system for load monitoring in horizontal-axis wind turbines. Choi et al. [126] determined the tip deflections of a composite wind turbine blade through a static load test using mechanical strains measured by FBG sensor probes.

Railway Infrastructure

Recently, the optical fiber sensor-based monitoring system has attracted great interests among the researchers in the fields of railway engineering and optical engineering. Yan et al. [127] proposed three FBG-based methods for strain measurement and axle counting in high-speed railway systems, and the advantages and limitations of these approaches were discussed in terms of feasibility and cost-effectiveness through laboratory verification and evaluation. Wei et al. [128, 129] described a real-time wheel defect detection system through deploying FBG sensors on rail tracks of the Hong Kong mass transit railway (MTR) to gain the track strains upon wheel-rail interaction and generate a reliable condition index reflecting the wheel condition, and the effectiveness of the introduced system was verified by extensive field tests. Filograno et al. [130] implemented an FBG-based railway security monitoring system on the Spanish high-speed line Madrid-Barcelona for train identification, axle counting, speed and acceleration detection, wheel imperfection monitoring, and dynamic load calculation.

Pimentel et al. [131] developed a hybrid fiber-optic/electrical train characterization system with a new weight-in-motion (WIM) algorithm for on-motion determination of the train speed, acceleration and weight distribution for traffic monitoring, and safety evaluation of a railway bridge in Portugal. Kerrouche et al. [132,133] conducted the research on structural monitoring of a decommissioned concrete railway bridge in Sweden loaded to failure by use of an FBG-based distributed sensor system.

Kang and Chung [134] developed an integrated FBG-based monitoring scheme for a maglev guideway in Korea to measure the parameters involving strains, curvatures, vertical defections, and frequencies which were compared with those obtained from the conventional sensors [135]. Yoon et al. [136] proposed a distributed Brillouin optical correlation domain analysis- (BOCDA-) based sensing system to measure the longitudinal strain distribution of a rail in real time, and the results of a spatial resolution of 3.8 cm and an accuracy of $\pm 15\,\mu\varepsilon$ were achieved under different loading conditions applied to a 2.8 m rail. Wang et al. [137] utilized A-thermal FBG sensors and electronic sensors to record performance changes in the continuous welded rail, and the monitoring results revealed that the optical fiber sensor was durable and capable of long-term monitoring and was capable of providing sensitive, clear, and stable signals.

Bocciolone et al. [138] presented the application of FBG sensors on a pantograph for monitoring of the contact force and the vertical acceleration of the pantograph head of the pantograph-catenary system in an underground line. Boffi et al. [139] developed an innovative optical fiber sensor-based system for online monitoring of the contact force between the pantograph and the catenary at low and high frequencies.

Geotechnical Structures

Regarding the applications in geotechnical engineering, Pei et al. [140, 141] developed an FBG-based in-place inclinometer for lateral displacement measurement of slopes in accordance with the classical indeterminate beam theory which was successfully installed in a slope in China for long-term displacement monitoring. Kister et al. [142] deployed FBG sensors in reinforced concrete foundation piles for strain and temperature monitoring and structural health condition assessment. Lu et al. [143] conducted the field measurement of the stress within the precast pile by use of the BOTDR-based optical fiber sensing technique. Kim et al. [144] developed a specially designed FBG-embedded tendon for the monitoring of the prestress force and load transfer of the ground anchor and the feasibility of the device was verified through laboratory and field tests. Legge et al. [145] developed an FBG-based stress cell for determination of the full state of three-dimensional stress at any accessible or predetermined point in a soil mass or structure.

CONCLUSIONS

This paper provides a summary of the research and development in the area of structural monitoring of civil infrastructure by use of the optical fiber sensing technology. Based on a comprehensive review of the optical fiber sensor-based theories, methods, technologies, and applications, the following concluding remarks are made: (i) due to their unique merits, the optical fiber sensors have been widely used in life-cycle monitoring of civil infrastructure such as bridges, buildings, tunnels, pipelines, wind turbines, railway infrastructure, and geotechnical structures; (ii) the optical fiber sensing technology is capable of measuring lots of types of measurands such as strains, temperatures, accelerations, deflections/displacements, cracks, and corrosion; and (iii) the exploitation of protection measures in sensor installation as well as the development of cost-effective optical fiber demodulation instruments are desirable in the further research.

ACKNOWLEDGMENTS

The work described in this paper was jointly supported by the National Science Foundation of China (Grant nos. 51308493 and U1234204), the Research Fund for the Doctoral Program of Higher Education of China (Grant no. 20130101120080), and the Fundamental Research Funds for the Central Universities of China (Grant no. 2013QNA4023).

REFERENCES

1.   Y. Q. Ni, X. W. Ye, and J. M. Ko, "Monitoring-based fatigue reliability assessment of steel bridges: analytical model and application," Journal of Structural Engineering, vol. 136, no. 12, pp. 1563–1573, 2010.

2.   Y. Q. Ni, X. W. Ye, and J. M. Ko, "Modeling of stress spectrum using long-term monitoring data and finite mixture distributions," Journal of Engineering Mechanics, vol. 138, no. 2, pp. 175–183, 2012.

3.   X. W. Ye, Y. Q. Ni, K. Y. Wong, and J. M. Ko, "Statistical analysis of stress spectra for fatigue life assessment of steel bridges with structural health monitoring data," Engineering Structures, vol. 45, pp. 166–176, 2012.

4.   X. W. Ye, Y. Q. Ni, T. T. Wai, K. Y. Wong, X. M. Zhang, and F. Xu, "A vision-based system for dynamic displacement measurement of long-span bridges: algorithm and verification," Smart Structures and Systems, vol. 12, no. 3-4, pp. 363–379, 2013. ·

5.   H. N. Li, T. H. Yi, L. Ren, D. S. Li, and L. S. Huo, "Reviews on innovations and applications in structural health monitoring for infrastructures," Structural Monitoring and Maintenance, vol. 1, no. 1, pp. 1–45, 2014.

6.   T. H. Yi, H. N. Li, and M. Gu, "Recent research and applications of GPS-based monitoring technology for high-rise structures," Structural Control and Health Monitoring, vol. 20, no. 5, pp. 649–670, 2013.

7.   T. H. Yi, H. N. Li, and H. M. Sun, "Multi-stage structural damage diagnosis method based on "energy-damage" theory," Smart Structures and Systems, vol. 12, no. 3-4, pp. 345–361, 2013.

8.   J. M. Nichols, S. T. Trickey, M. Seaver, and L. Moniz, "Use of fiber-optic strain sensors and holder exponents for detecting and localizing damage in an experimental plate structure," Journal of Intelligent Material Systems and Structures, vol. 18, no. 1, pp. 51–67, 2007.

9.   H. Murayama, K. Kageyama, K. Uzawa, K. Ohara, and H. Igawa, "Strain monitoring of a single-lap joint with embedded fiber-optic distributed sensors," Structural Health Monitoring, vol. 11, no. 3, pp. 325–344, 2011.

10.   R. A. Silva-Muñoz and R. A. Lopez-Anido, "Structural health monitoring of marine composite structural joints using

embedded fiber Bragg grating strain sensors," Composite Structures, vol. 89, no. 2, pp. 224–234, 2009.

11.   C. Pang, M. Yu, A. K. Gupta, and K. M. Bryden, "Investigation of smart multifunctional optical sensor platform and its application in optical sensor networks," Smart Structures and Systems, vol. 12, no. 1, pp. 23–39, 2013. ·

12.   A. Khiat, F. Lamarque, C. Prelle, P. Pouille, M. Leester-Schädel, and S. Büttgenbach, "Two-dimension fiber optic sensor for high-resolution and long-range linear measurements," Sensors and Actuators A: Physical, vol. 158, no. 1, pp. 43–50, 2010.

13.   F. Ansari, "State-of-the-art in the applications of fiber-optic sensors to cementitious composites,"Cement and Concrete Composites, vol. 19, no. 1, pp. 3–19, 1997.

14.   X. Bao and L. Chen, "Recent progress in Brillouin scattering based fiber sensors," Sensors, vol. 11, no. 4, pp. 4152–4187, 2011.

15.   K. T. V. Grattan and T. Sun, "Fiber optic sensor technology: an overview," Sensors and Actuators, A: Physical, vol. 82, no. 1, pp. 40–61, 2000.

16.   L. Deng and C. S. Cai, "Applications of fiber optic sensors in civil engineering," Structural Engineering and Mechanics, vol. 25, no. 5, pp. 577–596, 2007.

17.   H. N. Li, D. S. Li, and G. B. Song, "Recent applications of fiber optic sensors to health monitoring in civil engineering," Engineering Structures, vol. 26, no. 11, pp. 1647–1657, 2004.

18.   V. Bhatia, K. A. Murphy, R. O. Claus, T. A. Tran, and J. A. Greene, "Recent developments in optical-fiber-based extrinsic Fabry-Perot interferometric strain sensing technology," Smart Materials and Structures, vol. 4, no. 4, pp. 246–251, 1995.

19.   Y. J. Rao, "Recent progress in applications of in-fibre Bragg grating sensors," Optics and Lasers in Engineering, vol. 31, no. 4, pp. 297–324, 1999.

20.   C. K. Y. Leung, "Fiber optic sensors in concrete: the future?" NDT & E International, vol. 34, no. 2, pp. 85–94, 2001.

21. R. M. Measures, M. LeBlanc, K. Liu et al., "Fiber optic sensors for smart structures," Optics and Lasers in Engineering, vol. 16, no. 2-3, pp. 127–152, 1992.

22. C. I. Merzbacher, A. D. Kersey, and E. J. Friebele, "Fiber optic sensors in concrete structures: a review,"Smart Materials and Structures, vol. 5, no. 2, pp. 196–208, 1996.

23. J. M. López-Higuera, L. R. Cobo, A. Q. Incera, and A. Cobo, "Fiber optic sensors in structural health monitoring," Journal of Lightwave Technology, vol. 29, no. 4, pp. 587–608, 2011.

24. F. Ansari, "Practical implementation of optical fiber sensors in civil structural health monitoring,"Journal of Intelligent Material Systems and Structures, vol. 18, no. 8, pp. 879–889, 2007.

25. M. Majumder, T. K. Gangopadhyay, A. K. Chakraborty, K. Dasgupta, and D. K. Bhattacharya, "Fibre Bragg gratings in structural health monitoring—present status and applications," Sensors and Actuators A: Physical, vol. 147, no. 1, pp. 150–164, 2008.

26. Q. B. Li and F. Ansari, "Circumferential strain measurement of high strength concrete in triaxial compression by fiber optic sensor," International Journal of Solids and Structures, vol. 38, no. 42-43, pp. 7607–7625, 2001.

27. D. C. Lee, J. J. Lee, and I. B. Kwon, "Monitoring of fatigue crack growth in steel structures using intensity-based optical fiber sensors," Journal of Intelligent Material Systems and Structures, vol. 11, no. 2, pp. 100–107, 2000.

28. S. Jacobs, S. Matthys, G. De Roeck, L. Taerwe, W. de Waele, and J. Degrieck, "Testing of a prestressed concrete girder to study the enhanced performance of monitoring by integrating optical fiber sensors,"Journal of Structural Engineering, vol. 133, no. 4, pp. 541–549, 2007.

29. P. Moyo, J. M. W. Brownjohn, R. Suresh, and S. C. Tjin, "Development of fiber Bragg grating sensors for monitoring civil infrastructure," Engineering Structures, vol. 27, no. 12, pp. 1828–1834, 2005.

30. D. C. Betz, L. Staudigel, M. N. Trutzel, and M. Kehlenbach, "Structural monitoring using fiber-optic bragg grating sensors," Structural Health Monitoring, vol. 2, no. 2, pp. 145–152, 2003.

31. M. D. Todd, G. A. Johnson, and S. T. Vohra, "Deployment of a fiber bragg grating-based measurement system in a structural health monitoring application," Smart Materials and Structures, vol. 10, no. 3, pp. 534–539, 2001.

32. P. Capoluongo, C. Ambrosino, S. Campopiano et al., "Modal analysis and damage detection by Fiber Bragg grating sensors," Sensors and Actuators A: Physical, vol. 133, no. 2, pp. 415–424, 2007.

33. K. Kesavan, K. Ravisankar, S. Parivallal, and P. Sreeshylam, "Applications of fiber optic sensors for structural health monitoring," Smart Structures and Systems, vol. 1, no. 4, pp. 355–368, 2005.

34. I. B. Kwon, M. Y. Choi, and H. Moon, "Strain measurement using fiber optic total reflected extrinsic Fabry-Perot interferometric sensor with a digital signal processing algorithm," Sensors and Actuators A: Physical, vol. 112, no. 1, pp. 10–17, 2004.

35. T. Liu, D. Brooks, A. Martin, R. Badcock, B. Ralph, and G. F. Fernando, "A multi-mode extrinsic Fabry-Pérot interferometric strain sensor," Smart Materials and Structures, vol. 6, no. 4, pp. 464–469, 1997.

36. M. Z. Jiang and E. Gerhard, "Simple strain sensor using a thin film as a low-finesse fiber-optic Fabry-Perot interferometer," Sensors and Actuators A: Physical, vol. 88, no. 1, pp. 41–46, 2001.

37. F. Akhavan, S. E. Watkins, and K. Chandrashekhara, "Measurement and analysis of impact-induced strain using extrinsic Fabry-Pérot fiber optic sensors," Smart Materials and Structures, vol. 7, no. 6, pp. 745–751, 1998.

38. V. Bhatia, K. A. Murphy, R. O. Claus et al., "Multiple strain state measurements using conventional and absolute optical fiber-based extrinsic Fabry-Perot interferometric strain

sensors," Smart Materials and Structures, vol. 4, no. 4, pp. 240–245, 1995.

39. A. Güemes, A. Fernández-López, and B. Soller, "Optical fiber distributed sensing-physical principles and applications," Structural Health Monitoring, vol. 9, no. 3, pp. 233–245, 2010.

40. Z. Zhu, D. Liu, Q. Yuan, B. Liu, and J. Liu, "A novel distributed optic fiber transduser for landslides monitoring," Optics and Lasers in Engineering, vol. 49, no. 7, pp. 1019–1024, 2011.

41. H. Li, G. Zhou, L. Ren, and D. Li, "Strain transfer coefficient analyses for embedded fiber bragg grating sensors in different host materials," Journal of Engineering Mechanics, vol. 135, no. 12, pp. 1343–1353, 2009.

42. B. Yun, Y. Wang, A. Li, and Y. Cui, "Simulated annealing evolutionary algorithm for the fibre Bragg grating distributed strain sensor," Measurement Science and Technology, vol. 16, no. 12, pp. 2425–2430, 2005.

43. M. Imai and M. Feng, "Sensing optical fiber installation study for crack identification using a stimulated Brillouin-based strain sensor," Structural Health Monitoring, vol. 11, no. 5, pp. 501–509, 2012.

44. X. L. Zhang, D. K. Liang, J. Zeng, and A. Asundi, "Genetic algorithm-support vector regression for high reliability SHM system based on FBG sensor network," Optics and Lasers in Engineering, vol. 50, no. 2, pp. 148–153, 2012.

45. A. Gill, K. Peters, and M. Studer, "Genetic algorithm for the reconstruction of Bragg grating sensor strain profiles," Measurement Science and Technology, vol. 15, no. 9, pp. 1877–1884, 2004.

46. M. Prabhugoud and K. Peters, "Finite element model for embedded fiber Bragg grating sensor," Smart Materials and Structures, vol. 15, no. 2, pp. 550–562, 2006.

47. J. G. Liu, C. Schmidt-Hattenberger, and G. Borm, "Dynamic strain measurement with a fibre Bragg grating sensor system," Measurement, vol. 32, no. 2, pp. 151–161, 2002.

48.   Y. C. Ma, Y. H. Yang, J. M. Li, M. W. Yang, J. Tang, and T. Liang, "Dynamic and static strain gauge using superimposed fiber Bragg gratings," Measurement Science and Technology, vol. 23, no. 10, Article ID 105202, 2012.

49.   B. W. Jang, Y. G. Lee, J. H. Kim, Y. Y. Kim, and C. G. Kim, "Real-time impact identification algorithm for composite structures using fiber Bragg grating sensors," Structural Control and Health Monitoring, vol. 19, no. 7, pp. 580–591, 2012.

50.   C. Schizas, S. Stutz, J. Botsis, and D. Coric, "Monitoring of non-homogeneous strains in composites with embedded wavelength multiplexed fiber Bragg gratings: a methodological study," Composite Structures, vol. 94, no. 3, pp. 987–994, 2012.

51.   X. Feng, X. T. Zhang, C. S. Sun, M. Motamedi, and F. Ansari, "Stationary wavelet transform method for distributed detection of damage by fiber-optic sensors," Journal of Engineering Mechanics, vol. 140, no. 4, pp. 1–11, 2014.

52.   K. Peters, M. Studer, J. Botsis, A. Iocco, H. Limberger, and R. Salathé, "Embeded optical fiber bragg grating sensor in a nonuniform strain field: measurements and simulations," Experimental Mechanics, vol. 41, no. 1, pp. 19–28, 2001.

53.   Y. Lee, D. Kim, and C. Kim, "Performance of a single reflective grating-based fiber optic accelerometer,"Measurement Science and Technology, vol. 23, no. 4, pp. 1–7, 2012.

54.   Y. Wang and H. Huang, "Optical fiber corrosion sensor based on laser light reflection," Smart Materials and Structures, vol. 20, no. 8, Article ID 085003, pp. 1–7, 2011.

55.   L. Rodriguez-Cobo, A. T. Marques, J. M. Lopez-Higuera, J. L. Santos, and O. Frazao, "New design for temperature-strain discrimination using fiber Bragg gratings embedded in laminated composites,"Smart Materials and Structures, vol. 22, no. 10, pp. 1–10, 2013.

56.   S. Pirozzi, "Multi-point force sensor based on crossed optical fibers," Sensors and Actuators A: Physical, vol. 183, pp. 1–10, 2012.

57. D. H. Kim, J. W. Park, H. K. Kang, C. S. Hong, and C. G. Kim, "Measuring dynamic strain of structures using a gold-deposited extrinsic Fabry-Perot interferometer," Smart Materials and Structures, vol. 12, no. 1, pp. 1–5, 2003.

58. G. Cumunel, S. Delepine-Lesoille, and P. Argoul, "Long-gage optical fiber extensometers for dynamic evaluation of structures," Sensors and Actuators, A: Physical, vol. 184, pp. 1–15, 2012.

59. T. K. Gangopadhyay, M. Majumder, A. Kumar Chakraborty, A. Kumar Dikshit, and D. Kumar Bhattacharya, "Fibre Bragg grating strain sensor and study of its packaging material for use in critical analysis on steel structure," Sensors and Actuators A: Physical, vol. 150, no. 1, pp. 78–86, 2009.

60. L. B. Yuan, Q. B. Li, Y. J. Liang, J. Yang, and Z. H. Liu, "Fiber optic 2-D sensor for measuring the strain inside the concrete specimen," Sensors and Actuators A: Physical, vol. 94, no. 1, pp. 25–31, 2001.

61. H. Y. Liu, D. K. Liang, J. Zeng, J. Jin, J. Wu, and J. Geng, "Design of a long-period fiber grating sensor for reinforcing bar corrosion in concrete," Journal of Intelligent Material Systems and Structures, vol. 23, no. 1, pp. 45–51, 2012.

62. S. Yashiro, T. Okabe, and N. Takeda, "Damage identification in a holed CFRP laminate using a chirped fiber Bragg grating sensor," Composites Science and Technology, vol. 67, no. 2, pp. 286–295, 2007.

63. G. Zhou, L. M. Sim, and J. Loughlan, "Damage evaluation of smart composite beams using embedded extrinsic Fabry-Perot interferometric strain sensors: bending stiffness assessment," Smart Materials and Structures, vol. 13, no. 6, pp. 1291–1302, 2004.

64. S. Triollet, L. Robert, E. Marin, and Y. Ouerdane, "Discriminated measures of strain and temperature in metallic specimen with embedded superimposed long and short fibre Bragg gratings," Measurement Science and Technology, vol. 22, no. 1, Article ID 015202, 2011.

65. W. Hong, Z. S. Wu, C. Q. Yang, C. F. Wan, G. Wu, and Y. F. Zhang, "Condition assessment of reinforced concrete beams using dynamic data measured with distributed long-gage macro-strain sensors," Journal of Sound and Vibration, vol. 331, no. 12, pp. 2764–2782, 2012.

66. S. Li and Z. Wu, "Development of distributed long-gage fiber optic sensing system for structural health monitoring," Structural Health Monitoring, vol. 6, no. 2, pp. 133–143, 2007.

67. M. Quirion and G. Ballivy, "Concrete strain monitoring with Fabry-Perot fiber-optic sensor," Journal of Materials in Civil Engineering, vol. 12, no. 3, pp. 254–261, 2000.

68. C. E. Davis, P. Norman, C. Ratcliffe, and R. Crane, "Broad area damage detection in composites using fibre Bragg grating arrays," Structural Health Monitoring, vol. 11, no. 6, pp. 724–732, 2012.

69. S. Villalba and J. R. Casas, "Application of optical fiber distributed sensing to health monitoring of concrete structures," Mechanical Systems and Signal Processing, vol. 39, no. 1-2, pp. 441–451, 2013.

70. B. Torres, I. Payá-Zaforteza Ignacio, P. A. Calderón, and J. M. Adam, "Analysis of the strain transfer in a new FBG sensor for structural health monitoring," Engineering Structures, vol. 33, no. 2, pp. 539–548, 2011.

71. K. Schröder, W. Ecke, M. Kautz, S. Willett, M. Jenzer, and T. Bosselmann, "An approach to continuous on-site monitoring of contact forces in current collectors by a fiber optic sensing system," Optics and Lasers in Engineering, vol. 51, no. 2, pp. 172–179, 2013.

72. M. Kim, "A smart health monitoring system with application to welded structures using piezoceramic and fiber optic transducers," Journal of Intelligent Material Systems and Structures, vol. 17, no. 1, pp. 35–44, 2006.

73. J. C. Xu, G. Pickrell, X. W. Wang, W. Peng, K. Cooper, and A. Wang, "A novel temperature-insensitive optical fiber pressure

sensor for harsh environments," IEEE Photonics Technology Letters, vol. 17, no. 4, pp. 870–872, 2005.

74.   M. J. García, J. A. Ortega, J. A. Chávez, J. Salazar, and A. Turó, "A novel distributed fiber-optic strain sensor," IEEE Transactions on Instrumentation and Measurement, vol. 51, no. 4, pp. 685–690, 2002.

75.   T. Liu, M. Wu, Y. Rao, D. A. Jackson, and G. F. Fernando, "A multiplexed optical fibre-based extrinsic Fabry-Perot sensor system for in-situ strain monitoring in composites," Smart Materials and Structures, vol. 7, no. 4, pp. 550–556, 1998.

76.   Y. J. Sun, B. Shi, S. E. Chen, H. H. Zhu, D. Zhang, and Y. Lu, "Feasibility study on corrosion monitoring of a concrete column with central rebar using BOTDR," Smart Structures and Systems, vol. 13, no. 1, pp. 41–53, 2014.

77.   C. Lan, Z. Zhou, and J. Ou, "Monitoring of structural prestress loss in RC beams by inner distributed Brillouin and fiber Bragg grating sensors on a single optical fiber," Structural Control and Health Monitoring, vol. 21, no. 3, pp. 317–330, 2014.

78.   Z. Zhou, J. P. He, G. D. Chen, and J. P. Ou, "A smart steel strand for the evaluation of prestress loss distribution in post-tensioned concrete structures," Journal of Intelligent Material Systems and Structures, vol. 20, no. 16, pp. 1901–1912, 2009.

79.   J. R. Casas and P. J. S. Cruz, "Fiber optic sensors for bridge monitoring," Journal of Bridge Engineering, vol. 8, no. 6, pp. 362–373, 2003.

80.   E. Mehrani, A. Ayoub, and A. Ayoub, "Evaluation of fiber optic sensors for remote health monitoring of bridge structures," Materials and Structures, vol. 42, no. 2, pp. 183–199, 2009.

81.   B. Glisic and D. Inaudi, "Development of method for in-service crack detection based on distributed fiber optic sensors," Structural Health Monitoring, vol. 11, no. 2, pp. 161–171, 2012.

82.   I. Talebinejad, C. Fischer, and F. Ansari, "Serially multiplexed FBG accelerometer for structural health monitoring of

bridges," Smart Structures and Systems, vol. 5, no. 4, pp. 345–355, 2009.

83. R. C. Tennyson, A. A. Mufti, S. Rizkalla, G. Tadros, and B. Benmokrane, "Structural health monitoring of innovative bridges in Canada with fiber optic sensors," Smart Materials and Structures, vol. 10, no. 3, pp. 560–573, 2001.

84. R. Brönnimann, P. M. Nellen, and U. Sennhauser, "Application and reliability of a fiber optical surveillance system for a stay cable bridge," Smart Materials and Structures, vol. 7, no. 2, pp. 229–236, 1998.

85. C. Rodrigues, F. Cavadas, C. Félix, and J. Figueiras, "FBG based strain monitoring in the rehabilitation of a centenary metallic bridge," Engineering Structures, vol. 44, pp. 281–290, 2012.

86. C. Barbosa, N. Costa, L. A. Ferreira et al., "Weldable fibre Bragg grating sensors for steel bridge monitoring," Measurement Science and Technology, vol. 19, no. 12, Article ID 125305, 2008.

87. A. Kerrouche, W. J. O. Boyle, T. Sun, and K. T. V. Grattan, "Design and in-the-field performance evaluation of compact FBG sensor system for structural health monitoring applications," Sensors and Actuators, A: Physical, vol. 151, no. 2, pp. 107–112, 2009.

88. G. Kister, R. A. Badcock, Y. M. Gebremichael et al., "Monitoring of an all-composite bridge using Bragg grating sensors," Construction and Building Materials, vol. 21, no. 7, pp. 1599–1604, 2007.

89. G. Kister, D. Winter, R. A. Badcock et al., "Structural health monitoring of a composite bridge using Bragg grating sensors. Part 1: evaluation of adhesives and protection systems for the optical sensors,"Engineering Structures, vol. 29, no. 3, pp. 440–448, 2007.

90. M. R. Mokhtar, K. Owens, J. Kwasny et al., "Fiber-optic strain sensor system with temperature compensation for arch bridge

condition monitoring," IEEE Sensors Journal, vol. 12, no. 5, pp. 1470–1476, 2012.

91. F. Surre, T. Sun, and K. T. Grattan, "Fiber optic strain monitoring for long-term evaluation of a concrete footbridge under extended test conditionss," IEEE Sensors Journal, vol. 13, no. 3, pp. 1036–1043, 2013.

92. T. H. T. Chan, L. Yu, H. Y. Tam et al., "Fiber Bragg grating sensors for structural health monitoring of Tsing Ma bridge: background and experimental observation," Engineering Structures, vol. 28, no. 5, pp. 648–659, 2006.

93. M. Yau, T. Chan, D. Thambiratnam, and H. Tam, "Static vertical displacement measurement of bridges using fiber bragg grating (FBG) sensors," Advances in Structural Engineering, vol. 16, no. 1, pp. 165–176, 2013.

94. X. Zhao, J. Lu, R. C. Han, X. L. Kong, Y. H. Wang, and L. Li, "Application of multiscale fiber optical sensing network based on brillouin and fiber bragg grating sensing techniques on concrete structures,"International Journal of Distributed Sensor Networks, vol. 2012, Article ID 310797, 10 pages, 2012.

95. W. Chung, S. Kim, N. Kim, and H. Lee, "Deflection estimation of a full scale prestressed concrete girder using long-gauge fiber optic sensors," Construction and Building Materials, vol. 22, no. 3, pp. 394–401, 2008.

96. Y. B. Lin, C. L. Pan, Y. H. Kuo, K. C. Chang, and J. C. Chern, "Online monitoring of highway bridge construction using fiber Bragg grating sensors," Smart Materials and Structures, vol. 14, no. 5, pp. 1075–1082, 2005.

97. C. Rodrigues, C. Félix, A. Lage, and J. Figueiras, "Development of a long-term monitoring system based on FBG sensors applied to concrete bridges," Engineering Structures, vol. 32, no. 8, pp. 1993–2002, 2010.

98. G. Jiang, M. Dawood, K. Peters, and S. Rizkalla, "Global and local fiber optic sensors for health monitoring of civil

engineering infrastructure retrofit with FRP materials," Structural Health Monitoring, vol. 9, no. 4, pp. 309–322, 2010.

99. W. Zhang, J. Gao, B. Shi, H. Cui, and H. Zhu, "Health monitoring of rehabilitated concrete bridges using distributed optical fiber sensing," Computer-Aided Civil and Infrastructure Engineering, vol. 21, no. 6, pp. 411–424, 2006.

100. B. J. A. Costa and J. A. Figueiras, "Fiber optic based monitoring system applied to a centenary metallic arch bridge: design and installation," Engineering Structures, vol. 44, pp. 271–280, 2012.

101. J. P. He, Z. Zhou, and O. Jinping, "Optic fiber sensor-based smart bridge cable with functionality of self-sensing," Mechanical Systems and Signal Processing, vol. 35, no. 1-2, pp. 84–94, 2013.

102. H. Li, J. Ou, and Z. Zhou, "Applications of optical fibre Bragg gratings sensing technology-based smart stay cables," Optics and Lasers in Engineering, vol. 47, no. 10, pp. 1077–1084, 2009.

103. D. Li, Z. Zhou, and J. Ou, "Development and sensing properties study of FRP-FBG smart stay cable for bridge health monitoring applications," Measurement, vol. 44, no. 4, pp. 722–729, 2011.

104. D. S. Li, Z. Zhou, and J. P. Ou, "Dynamic behavior monitoring and damage evaluation for arch bridge suspender using GFRP optical fiber Bragg grating sensors," Optics and Laser Technology, vol. 44, no. 4, pp. 1031–1038, 2012.

105. Z. Zhou, M. H. Huang, L. Q. Huang, J. P. Ou, and G. D. Chen, "An optical fiber bragg grating sensing system for scour monitoring," Advances in Structural Engineering, vol. 14, no. 1, pp. 67–78, 2011.

106. Y. Lin, J. Chen, K. Chang, J. Chern, and J. Lai, "Real-time monitoring of local scour by using fiber Bragg grating sensors," Smart Materials and Structures, vol. 14, no. 4, pp. 664–670, 2005.

107. Y. B. Lin, J. S. Lai, K. C. Chang, and L. S. Li, "Flood scour monitoring system using fiber Bragg grating sensors," Smart Materials and Structures, vol. 15, no. 6, pp. 1950–1959, 2006.

108. W. Xiong, C. S. Cai, and X. Kong, "Instrumentation design for bridge scour monitoring using fiber Bragg grating sensors," Applied Optics, vol. 51, no. 5, pp. 547–557, 2012.

109. F. Bastianini, M. Corradi, A. Borri, and A. D. Tommaso, "Retrofit and monitoring of an historical building using "smart" CFRP with embedded fibre optic Brillouin sensors," Construction and Building Materials, vol. 19, no. 7, pp. 525–535, 2005.

110. P. Antunes, R. Travanca, H. Rodrigues et al., "Dynamic structural health monitoring of slender structures using optical sensors," Sensors, vol. 12, no. 5, pp. 6629–6644, 2012.

111. Y. Q. Ni, Y. Xia, W. Y. Liao, and J. M. Ko, "Technology innovation in developing the structural health monitoring system for Guangzhou New TV Tower," Structural Control and Health Monitoring, vol. 16, no. 1, pp. 73–98, 2009.

112. D. S. Li, L. Ren, H. N. Li, and G. B. Song, "Structural health monitoring of a tall building during construction with fiber Bragg grating sensors," International Journal of Distributed Sensor Networks, vol. 2012, Article ID 272190, 10 pages, 2012.

113. X. W. Ye, Y. Q. Ni, and J. H. Yin, "Safety monitoring of railway tunnel construction using FBG sensing technology," Advances in Structural Engineering, vol. 16, no. 8, pp. 1401–1409, 2013.

114. N. Metje, D. N. Chapman, C. D. F. Rogers, P. Henderson, and M. Beth, "An optical fiber sensor system for remote displacement monitoring of structures—prototype tests in the laboratory," Structural Health Monitoring, vol. 7, no. 1, pp. 51–63, 2008.

115. C. Li, Y. Zhao, H. Liu, Z. Wan, C. Zhang, and N. Rong, "Monitoring second lining of tunnel with mounted fiber Bragg

grating strain sensors," Automation in Construction, vol. 17, no. 5, pp. 641–644, 2008.

116. C. Li, Y. Zhao, H. Liu et al., "Strain and back cavity of tunnel engineering surveyed by FBG strain sensors and geological radar," Journal of Intelligent Material Systems and Structures, vol. 20, no. 18, pp. 2285–2289, 2009.

117. B. Glisic and Y. Yao, "Fiber optic method for health assessment of pipelines subjected to earthquake-induced ground movement," Structural Health Monitoring, vol. 11, no. 6, pp. 696–711, 2012.

118. C. Zhang, X. Bao, I. F. Ozkan et al., "Prediction of the pipe buckling by using broadening factor with distributed Brillouin fiber sensors," Optical Fiber Technology, vol. 14, no. 2, pp. 109–113, 2008.

119. L. Glavind, I. S. Olesen, B. F. Skipper, and M. Kristensen, "Fiber-optical grating sensors for wind turbine blades: a review," Optical Engineering, vol. 52, no. 3, Article ID 030901, pp. 1–9, 2013.

120. T. J. Arsenault, A. Achuthan, P. Marzocca, C. Grappasonni, and G. Coppotelli, "Development of a FBG based distributed strain sensor system for wind turbine structural health monitoring," Smart Materials and Structures, vol. 22, no. 7, Article ID 075027, 2013.

121. S. W. Kim, W. R. Kang, M. S. Jeong, I. Lee, and I. B. Kwon, "Deflection estimation of a wind turbine blade using FBG sensors embedded in the blade bonding line," Smart Materials and Structures, vol. 22, no. 12, pp. 1–11, 2013.

122. J. Burgmeier, W. Schippers, N. Emde, P. Funken, and W. Schade, "Femtosecond laser-inscribed fiber Bragg gratings for strain monitoring in power cables of offshore wind turbines," Applied Optics, vol. 50, no. 13, pp. 1868–1872, 2011.

123. H. J. Bang, H. I. Kim, and K. S. Lee, "Measurement of strain and bending deflection of a wind turbine tower using arrayed FBG sensors," International Journal of Precision Engineering and Manufacturing, vol. 13, no. 12, pp. 2121–2126, 2012.

124. Y. Ge, K. S. Kuang, and S. T. Quek, "Development of a low-cost bi-axial intensity-based optical fibre accelerometer for wind turbine blades," Sensors and Actuators A: Physical, vol. 197, pp. 126–135, 2013.

125. K. Schroeder, W. Ecke, J. Apitz, E. Lembke, and G. Lenschow, "A fibre Bragg grating sensor system monitors operational load in a wind turbine rotor blade," Measurement Science and Technology, vol. 17, no. 5, pp. 1167–1172, 2006.

126. K. S. Choi, Y. H. Huh, I. B. Kwon, and D. J. Yoon, "A tip deflection calculation method for a wind turbine blade using temperature compensated FBG sensors," Smart Materials and Structures, vol. 21, no. 2, Article ID 025008, pp. 1–9, 2012.

127. L. S. Yan, Z. T. Zhang, P. Wang et al., "Fiber sensors for strain measurements and axle counting in high-speed railway applications," IEEE Sensors Journal, vol. 11, no. 7, pp. 1587–1594, 2011.

128. C. Wei, C. Lai, S. Liu et al., "A fiber Bragg grating sensor system for train axle counting," IEEE Sensors Journal, vol. 10, no. 12, pp. 1905–1912, 2010.

129. C. Wei, Q. Xin, W. H. Chung, S. Liu, H. Tam, and S. L. Ho, "Real-time train wheel condition monitoring by fiber Bragg grating sensors," International Journal of Distributed Sensor Networks, vol. 2012, Article ID 409048, 7 pages, 2012.

130. M. L. Filograno, P. C. Guillen, A. Rodriguez-Barrios et al., "Real-time monitoring of railway traffic using fiber Bragg grating sensors," IEEE Sensors Journal, vol. 12, no. 1, pp. 85–92, 2012.

131. R. M. C. M. Pimentel, M. C. B. Barbosa, N. M. S. Costa et al., "Hybrid fiber-optic/electrical measurement system for characterization of railway traffic and its effects on a short span bridge," IEEE Sensors Journal, vol. 8, no. 7, pp. 1243–1249, 2008.

132. A. Kerrouche, J. Leighton, W. J. O. Boyle et al., "Strain measurement on a rail bridge loaded to failure using a fiber

Bragg grating-based distributed sensor system," IEEE Sensors Journal, vol. 8, no. 12, pp. 2059–2065, 2008.

133. A. Kerrouche, W. J. O. Boyle, Y. M. Gebremichael et al., "Field tests of fibre Bragg grating sensors incorporated into CFRP for railway bridge strengthening condition monitoring," Sensors and Actuators A: Physical, vol. 148, no. 1, pp. 68–74, 2008.

134. D. Kang and W. Chung, "Integrated monitoring scheme for a maglev guideway using multiplexed FBG sensor arrays," NDT and E International, vol. 42, no. 4, pp. 260–266, 2009.

135. W. Chung and D. Kang, "Full-scale test of a concrete box girder using FBG sensing system," Engineering Structures, vol. 30, no. 3, pp. 643–652, 2008.

136. H. J. Yoon, K. Y. Song, J. S. Kim, and D. S. Kim, "Longitudinal strain monitoring of rail using a distributed fiber sensor based on Brillouin optical correlation domain analysis," NDT and E International, vol. 44, no. 7, pp. 637–644, 2011.

137. C. Wang, H. Tsai, C. Chen, and H. Wang, "Railway track performance monitoring and safety warning system," Journal of Performance of Constructed Facilities, vol. 25, no. 6, pp. 577–586, 2011.

138. M. Bocciolone, G. Bucca, A. Collina, and L. Comolli, "Pantograph-catenary monitoring by means of fibre Bragg grating sensors: results from tests in an underground line," Mechanical Systems and Signal Processing, vol. 41, no. 1-2, pp. 226–238, 2013.

139. P. Boffi, G. Cattaneo, L. Amoriello et al., "Optical fiber sensors to measure collector performance in the pantograph-catenary interaction," IEEE Sensors Journal, vol. 9, no. 6, pp. 635–640, 2009.

140. H. Pei, J. Yin, H. Zhu, C. Hong, W. Jin, and D. Xu, "Monitoring of lateral displacements of a slope using a series of special fibre Bragg grating-based in-place inclinometers," Measurement Science and Technology, vol. 23, no. 2, Article ID 025007, 2012.

141. H. Pei, J. Yin, and W. Jin, "Development of novel optical fiber sensors for measuring tilts and displacements of geotechnical structures," Measurement Science and Technology, vol. 24, no. 9, Article ID 095202, 2013.

142. G. Kister, D. Winter, Y. M. Gebremichael et al., "Methodology and integrity monitoring of foundation concrete piles using Bragg grating optical fibre sensors," Engineering Structures, vol. 29, no. 9, pp. 2048–2055, 2007.

143. Y. Lu, B. Shi, G. Q. Wei, S. E. Chen, and D. Zhang, "Application of a distributed optical fiber sensing technique in monitoring the stress of precast piles," Smart Materials and Structures, vol. 21, no. 11, Article ID 115011, 2012.

144. Y. S. Kim, H. J. Sung, H. W. Kim, and J. M. Kim, "Monitoring of tension force and load transfer of ground anchor by using optical FBG sensors embedded tendon," Smart Structures and Systems, vol. 7, no. 4, pp. 303–317, 2011.

145. T. F. H. Legge, P. L. Swart, G. van Zyl, and A. A. Chtcherbakov, "A fibre Bragg grating stress cell for geotechnical engineering applications," Measurement Science and Technology, vol. 17, no. 5, pp. 1173–1179, 2006.

Chapter 3

Environmental Degradation and Durability of Epoxy-Clay Nanocomposites

Raman P. Singh[1], Mikhail Khait[2], Suraj C. Zunjarrao[1], Chad S. Korach[2], and Gajendra Pandey[1]

[1]Mechanics of Advanced Materials Laboratory, School of Mechanical and Aerospace Engineering, Oklahoma State University, 218 Helmerich Research Center, 700 N. Greenwood Avenue, Tulsa, OK 74106-0700, USA

[2]Laboratory for Nanotribology and Wear Mechanics, Department of Mechanical Engineering, State University of New York, 131 Light Engineering, Stony Brook, NY 11794-2300, USA

ABSTRACT

This experimental investigation reports on the durability of epoxy-clay nanocomposites upon exposure to multiple environments.

Nanocomposites are fabricated by mixing the clay particles using various combinations of mechanical mixing, high-shear dispersion, and ultrasonication. Clay morphology is characterized using X-ray diffraction and transmission electron microscopy. Specimens of both neat epoxy and the epoxy-clay nanocomposite are subjected to two environmental conditions: combined UV radiation and condensation on 3-hour repeat cycle and constant temperature-humidity, for a total exposure duration of 4770 hours. The presence of nanoscale clay inhibits moisture uptake, as demonstrated by exposure to constant temperature-humidity. Nonetheless, both materials lose mass under exposure to combined UV radiation and condensation due to the erosion of epoxy by a synergistic process. Surprisingly, the epoxy-clay specimens exhibit greater mass loss, as compared to neat epoxy. Mechanical testing shows that either environment does not significant affect the flexure modulus of either material. On the other hand, both materials undergo degradation in flexural strength when exposed to either environment. However, the epoxy-clay nanocomposite retains 37% more flexure strength than the neat epoxy after 4072 hours of exposure.

INTRODUCTION

Epoxy-based thermosetting polymer resins have received great attention for aerospace and automotive applications due to their superior thermal and mechanical properties such as high modulus, high creep resistance, high glass transition temperature, low shrinkage at elevated temperature, and good resistance to chemicals. These properties have motivated the use of these resins as the matrix in fiber-reinforced composites that are currently being used extensively in aerospace and automotive industries. Such composites have also found applications in infrastructure applications such as bridges, buildings, tranport pipelines, and off-shore drilling platforms due to their superior and directional dependent properties. Despite their inherent advantages, however, such composites are susceptible to environmental conditions, primarily due to the degradation of the epoxy matrix. Hence, there are concerns regarding their overall long-term durability, especially as related to the capacity for sustained performance under harsh and changing environmental conditions.

The fiber reinforced composites used in infrastructure applications undergo mechanical and thermal loading while being exposed to adverse environments including ultraviolet (UV) radiation, moisture (relative humidity), water vapor condensation, and alkaline/salt environment [1]. The properties of the composites degrade when subjected to these harsh environments. Compared to other environments, UV radiation and water vapor are considered to be predominantly responsible for degradation during outdoor service [2]. Since most polymers have bond dissociation energy in the range of the wavelength of UV radiation (290–400 nm), they get affected greatly by exposure to the solar spectrum. Ultraviolet photons absorbed by polymers result in photo-oxidative reactions that alter the chemical structure by molecular chain scission or chain crosslinking that results in material deterioration [3]. Chain scission lowers the molecular weight of the polymer, giving rise to reduced strength and heat resistance; chain crosslinking leads to excessive brittleness and can result in microcracking. For prolonged exposure to UV radiation, the matrix dominated properties, such as interlaminar shear strength, flexural strength, and flexural stiffness can suffer severe deterioration [2, 4–7]. Moisture diffusion into the epoxy matrix on the other hand leads to plasticization and hydrolysis, which can cause both reversible and irreversible changes in thermophysical, mechanical, and chemical characteristics [1,7–9]. Both these processes lower the modulus, glass transition temperature and other matrix-dominated properties such as compressive strength, interlaminar shear strength; fatigue and impact tolerance can also deteriorate [1, 9–14]. Plasticization is usually reversible upon desorption of moisture, while hydrolysis of chemical bonds results in permanent irreversible damage. At the same time, moisture wicking along the fiber-matrix interface can degrade the fiber-matrix bond, resulting in loss of microstructural integrity. Furthermore, degradation phenomena due to UV radiation and moisture when acting together can significantly accelerate the degradation process of the matrix. Kumar et al. [6] studied the combined effect of UV and water vapor condensation and found that cyclic exposure leads to a synergistic degradation mechanism causing extensive matrix erosion and resulting loss of mechanical properties.

The reinforcement of polymers with nanoscale fillers offers the potential for significantly enhanced mechanical [15], thermal, and barrier properties [16]. In these nanocomposites, the enhancement in properties is directly related to the surface area of the reinforcement [17]. For a given volume fraction, the surface area of these nanoscale fillers is much higher as compared to that of micron-sized fillers. As a result, significant enhancement in various properties can be observed using very low volume fractions of nanoscale fillers. Among the various nano-sized reinforcements, high-aspect ratio layered silicates (nano-clay) are especially attractive for enhancing the barrier properties and hence the resistance to environmental degradation. Furthermore, nano-clays are readily available, are cheaper than other nanoscale fillers, and have well understood intercalation chemistry [17–19]. Polymer-clay nanocomposites can exhibit markedly improved mechanical, thermal, and physicochemical properties, as first demonstrated by Toyota researchers in the early 1990s [20]. Motivated by these studies, many researchers have synthesized and studied the properties of various clay-filled thermoplastic resins such as polyamides, polyimides, polyethylene, poly(methyl methacrylate), and ethylene vinyl acetate copolymers. Alexandre and Dubois [21] provide a summary of different polymers and fabrication routes available in the literature. The focus on polymer-clay nanocomposites based on thermosetting resins, especially epoxy, is more recent [18, 22–24].

In comparison to spherical or fiber-like reinforcement the addition of nanoscale clay reinforcement more effectively enhances the barrier properties such as resistance to helium permeability [25] and resistance to moisture transport [26–29]. Kim et al. [29] found that the appropriate choice of organically modified clay can reduce the moisture permeability by as much as 80% for clay loading of only 4 wt.%. Furthermore, Kim et al. observe much higher glass transition temperature for the nanocomposites than the neat epoxy under both dry and wet conditions. These results indicate that reduced moisture permeability could then lead to enhanced environmental resistance and that hypothesis motivates this investigation.

The focus of this paper is on the characterization of the environmental degradation of epoxy-clay nanocomposite under cyclic exposure to both ultraviolet radiation and water vapor condensation. In the past, our studies have shown that these two environments operate in a synergistic manner that leads to extensive degradation by the erosion of the epoxy matrix [6]. It is expected that the addition of clay could mitigate this mechanism by offering a barrier to moisture transport. Furthermore, for the mixing of clay particles in epoxy, prolonged mechanical mixing in association with high-shear dispersion or ultrasonication was investigated following another previous study [24]. The effect of these mixing techniques on the clay morphology and in turn on mechanical properties of epoxy-clay nanocomposites is also investigated.

EXPERIMENTAL METHODS

Material Fabrication

A commercially available octadecyl ammonium ion modified montmorillonite (MMT) layered silicate (Nanomer I.30E, Nanocor Inc., Arlington Heights, Illinois, USA) was used as the reinforcing agent. The epoxy resin used was Epon 862 (Shell Chemicals, USA) which is a diglycidyl ether of bisphenol F, and the curing agent used was Epikure 3274 (Shell Chemicals, USA) which is a moderately reactive, low viscosity aliphatic amine curing agent. A mechanical mixer (Model 350 Lab Stirrer, Arrow Engineering, Hillside, New Jersey, USA), a high speed shear dispenser (SV 25 KV—25 F dispersing element, IKA Works Inc., Wilmington,

North Carolina, USA), and an ultrasonic agitator (Model TSX750, Tekmar Dohrmann, Mason, Ohio, USA) were used to mix the clay in epoxy resin in five different ways.

Two different set of samples were fabricated in order to first establish the mixing process by enhancing the mechanical properties, then characterizing the environmental durability of the epoxy-clay nanocomposite. In the first set of samples 2% clay by volume fraction was used, and the mechanical properties of the different nanocomposite fabricated by using different mixing techniques were compared. To maximize the barrier properties due to addition of clay, it is desired to maximize the volume fraction of the same in the epoxy resin. Previous work by Zunjarrao et al. has shown that no agglomeration-induced deterioration of mechanical properties occurs as the volume fraction of clay is increased up to 4% [24]. Therefore, 4 vol.% clay was used to fabricate epoxy-clay nanocomposite by using the effecting mixing technique that was established before to evaluate the environmental durability of the nanocomposite. Note that, mechanical property enhancement, especially fracture toughness, is directly correlated to exfoliation and dispersion of clay in the epoxy matrix. The greater the exfoliation, the more the enhancement in barrier properties, and thus, greater resistance to environmental degradation.

The first set of epoxy-clay nanocomposite fabrication started with the addition of 2% clay by volume fraction to liquid epoxy resin preheated to a temperature of 60°C. The clay was then dispersed in epoxy by five different ways generating five nanocomposites with same constituents but formed by different processing techniques. In the first and second cases clay was dispersed by simply ultrasonication or high-shear dispersion, respectively, both for a period of 30 minutes. Both these processes led to heat generation due to viscous mixing. Since high temperatures can have a deleterious affect on the epoxy resin used, the temperature of the mixture was maintained at 60–65°C by using an ice bath. In the third case, clay was dispersed by stirring the mixture using a mechanical stirrer for a period of 14 hours. While in the fourth and fifth cases, the epoxy-clay mixture was first mixed for 14 hours using a mechanical stirrer and then further mixed using ultrasonication and high-shear dispersion, respectively, both again for 30 minutes. All the five mixtures were further processed and subjected to in situ polymerization using the same procedure. Mixture of clay and epoxy was degassed in a vacuum chamber for 12 hours to completely remove trapped air. A stoichiometric amount of the curing agent was added and hand mixed gently to avoid introduction of any air bubbles due to mixing action. The final slurry, free of air bubbles, was poured into an aluminium mold and allowed to cure under room temperature for 24 hours followed by postcuring at 121°C for 6 hours. Specimens for fracture toughness and flexural modulus tests were then cut out of the final cured sheet.

Two types of materials were prepared for environmental durability testing, Neat epoxy samples were prepared as before using Epon 862 and Epikure 3274 mixed with mechanical agitation. The mixed resin was degassed under vacuum for one hour before being poured into a teflon-coated aluminum mold. Curing occurred at 2C for 48 hours followed by postcuring at 121°C for 8 hours. The epoxy-clay nanocomposite was fabricated using the previously established procedure that results in optimal dispersion of the nanoclay. First, a desired amount of clay, to result in a final volume fraction of 4%, was added to liquid epoxy resin preheated to 60°C. This mixture was mixed using a mechanical stirrer for 14 hours while maintaining the temperature at 60°C using a hot plate. Subsequently, the mixture was processed for 30 minutes in a high-speed shear dispenser operating at 15,000 rpm. The high-shear process produces heat due to frictional dissipation in the polymer, and the temperature of the mixture was maintained at 60–65°C using an ice bath. The mixture was then degassed in a vacuum chamber for 2 hours to completely remove any trapped air. A stoichiometric amount of the curing agent was then added and hand mixed gently to avoid introduction of any air bubbles due to mixing action. The final slurry, free of air bubbles, was poured into a teflon-coated aluminum mold and allowed to cure at 25°C for 48 hours followed by postcuring at 121°C for 8 hours. In this manner, three sheets each of neat epoxy and epoxy-clay

Mass Loss or Gain upon Environmental Exposure

When exposed to constant relative humidity conditions in the temperature-humidity chamber, both neat epoxy and epoxy-clay composite samples gained mass due to moisture uptake, as can be seen in Figures 3 and4. These figures show data from multiple samples, to demonstrate scatter, and also plots the averaged variation. A comparison of the moisture absorption for neat epoxy and the epoxy-clay nanocomposites is shown in Figure 5 after taking an average reading from all the samples.

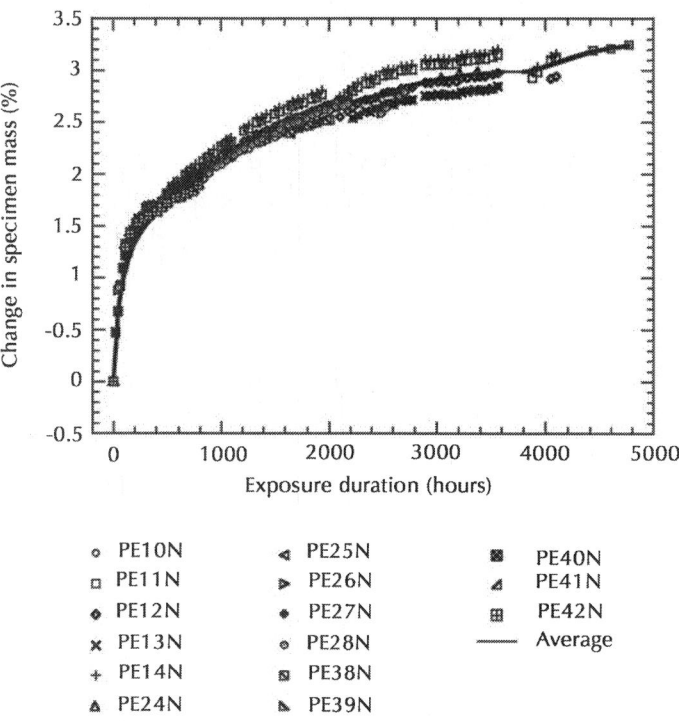

Figure 3: Moisture uptake for neat epoxy samples upon exposure to at moisture 80% relative humidity and 50°C.

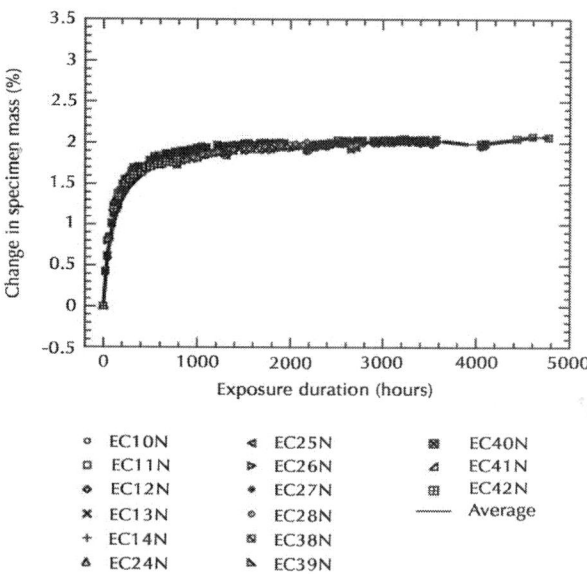

Figure 4: Moisture uptake for epoxy-clay composite samples upon exposure to at moisture 80% relative humidity and 50°C.

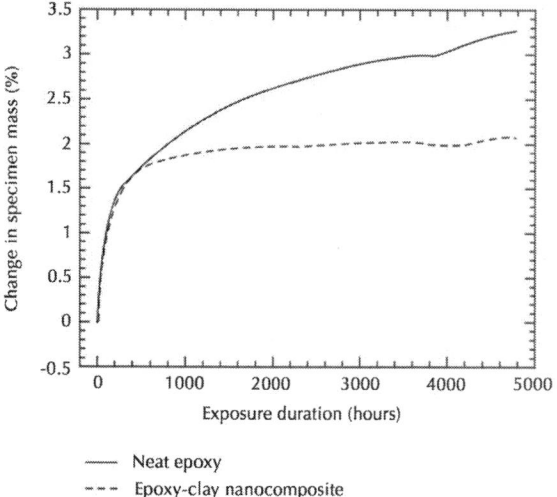

Figure 5: Comparison of average moisture uptake for neat epoxy and epoxy-clay nanocomposite specimens.

Both materials started gaining mass at a similar high rate for the first 250–325 hours and exhibited an increase of 1.6%. Shortly thereafter, the rate of moisture absorption for the epoxy-clay nanocomposite was minimal. Meanwhile, the neat epoxy specimens continued to increase in mass, and were not approaching saturation even after 4770 hours of exposure. The final mass gain by the epoxy-clay samples was 2.0% and for neat epoxy was 3.4%.

The variations in specimen mass for neat epoxy and epoxy-clay nanocomposite specimens exposed to combined UV radiation and condensation are shown in Figures 6 and 7. A comparison of averaged mass loss data for neat epoxy and epoxy-clay nanocomposites exposed to combined UV radiation and condensation is shown in Figure 8. In both cases, the samples gained in mass, for the first 100–500 hours, due to moisture uptake by the epoxy. Subsequently, the specimens started to lose mass by an epoxy erosion process. This process was first demonstrated by Kumar et al. [6], albeit for a different epoxy matrix, and is related to the synergistic physicochemical degradation of epoxy upon exposure to combined UV radiation and condensation.

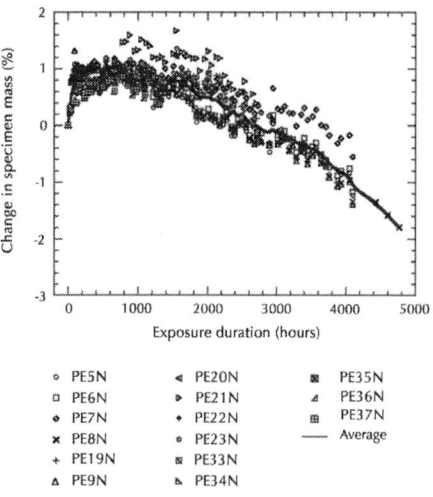

Figure 6: Mass variation for neat epoxy specimens upon exposure to combined UV radiation and water vapor condensation.

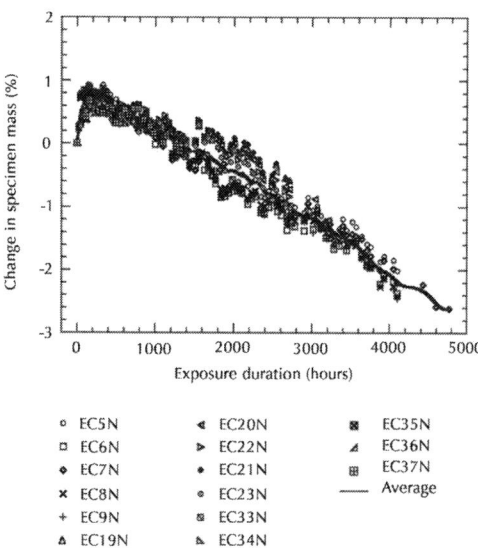

Figure 7: Mass variation for epoxy-clay nanocomposite specimens upon exposure to combined UV radiation and water vapor condensation.

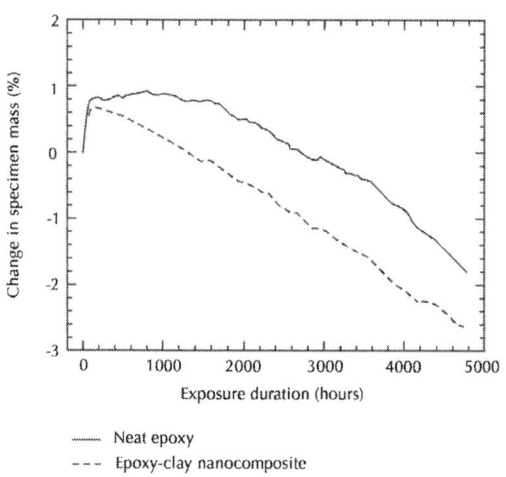

Figure 8: Comparison of mass variation for neat epoxy and epoxy-clay nanocomposite specimens upon exposure to combined UV radiation and water vapor condensation.

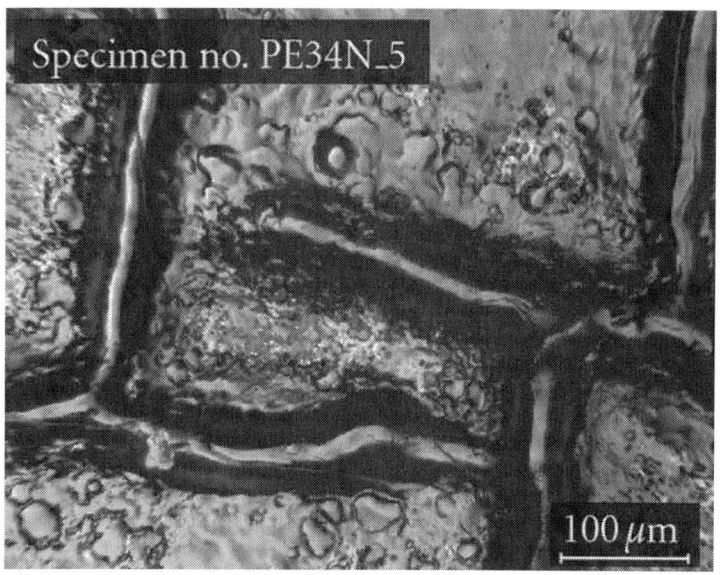

(a)

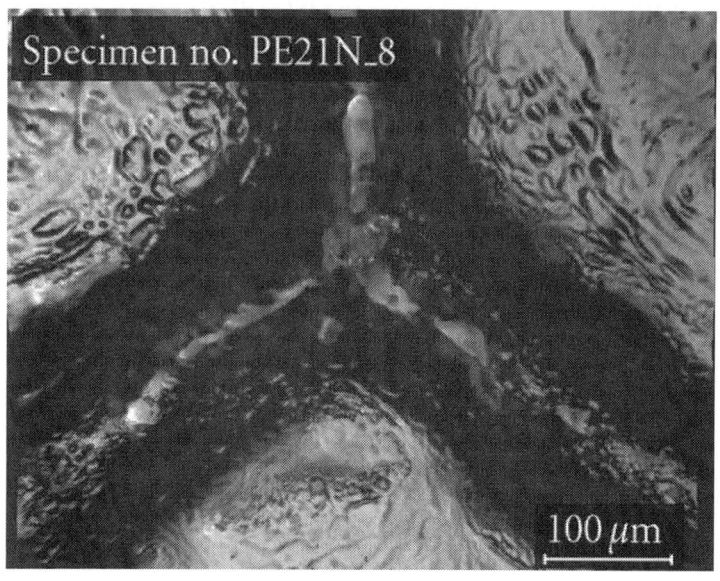

(b)

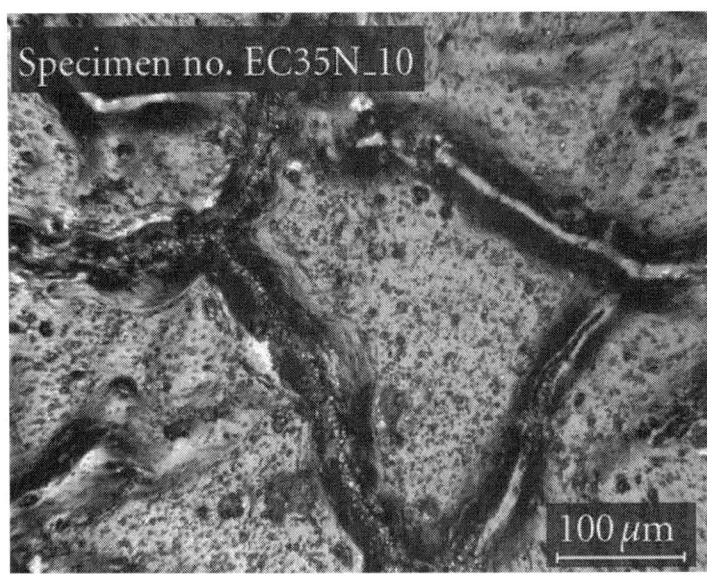

(c)

Figure 9: Images of neat epoxy specimens after exposure to UV radiation and condensation showing progressive erosion of microcracks.

A similar set of images obtained for epoxy-clay composite samples is shown in Figure 10. In this case, as for the neat epoxy, material degradation proceeded by erosion of channels initially formed due to UV radiation-induced microcracking. Also, in addition to the increase in size and depth of the trenches (channels), the density of trenches increased as well. This is simply due to the formation of additional microcracks during the UV radiation segments of the cyclic exposure protocol.

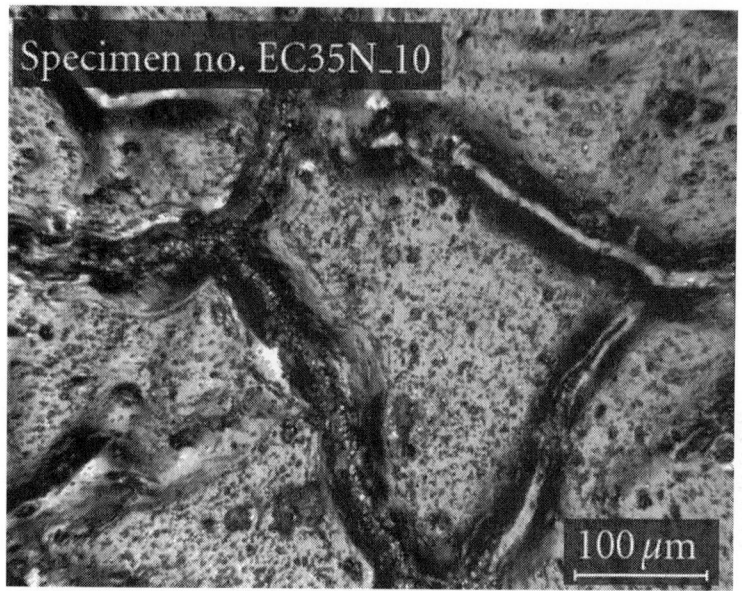

(a)

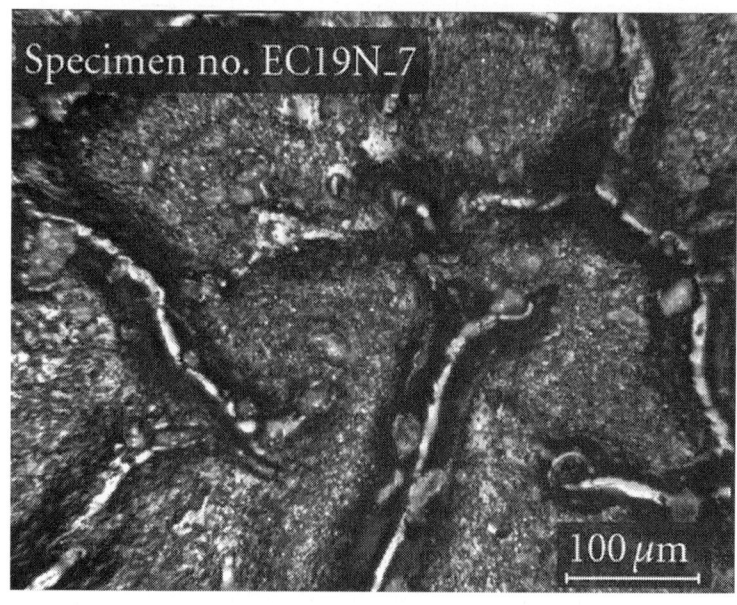

(b)

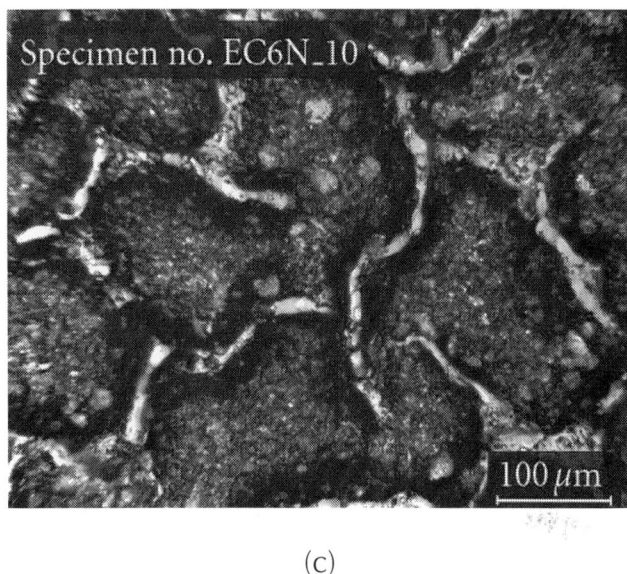

(c)

Figure 10: Images of epoxy-clay composites after exposure to UV radiation and condensation showing progressive erosion of microcracks.

From a direct visual comparison of Figures 9 and 10 it is apparent that less material erosion takes place for the epoxy-clay nanocomposites. This indicates that the presence of clay platelets helps in ameliorating material loss due to synergistic material degradation. Figures 11 and 12 show the evolution of trench cross-section profiles, as a function of exposure duration, for the neat epoxy and the epoxy-clay composite samples, respectively. Data for both materials has been plotted at the same scale to facilitate comparison. From the figures it is apparent that trench formation in the case of neat epoxy was much more aggressive than for the case of epoxy-clay composites. This again indicates that the presence of clay is significant in slowing down material erosion due to synergistic degradation. It is important to note that Figures 11 and 12 are not representative of the entire surface and are localized areas of degradation; other locations on the same sample would yield different crack depths and widths. Nonetheless, these figures illustrate the quantitative difference in trench formation for the two materials.

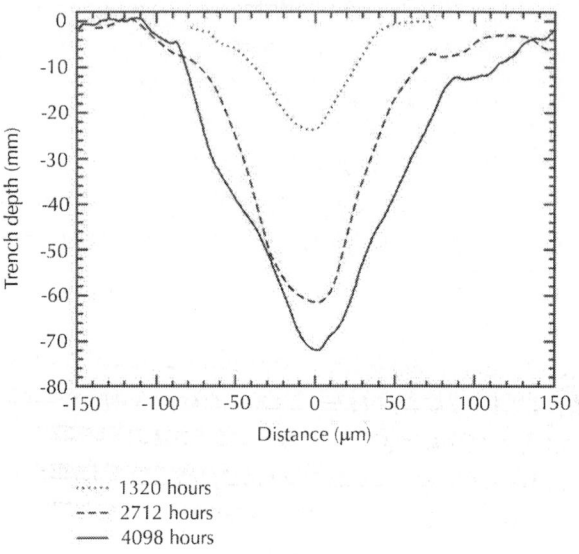

Figure 11: Selected trench profiles for neat epoxy samples at various stages of environmental degradation.

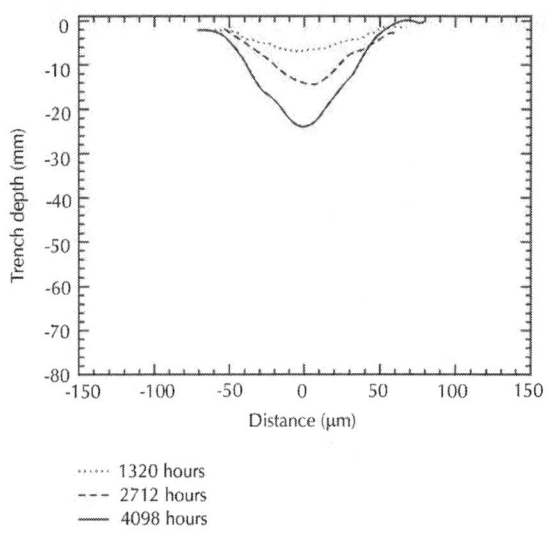

Figure 12: Selected trench profiles for epoxy-clay nanocomposite samples at various stages of environmental degradation.

The 3-dimensional imaging data was also used to determine the changes in volume and surface area for the two materials. The microscope cannot directly provide the actual change in volume due to material degradation. Instead, it provides a measurement of the current volume of the remaining material. This data was used to calculate the relative change in material volume, as a function of exposure duration. The change in material volume, normalized with respect to the viewing area of the Keyence microscope, is plotted in Figure 13. The average rate of volume change for the neat epoxy materials was ~6.4 ×10^{-3} $\mu m^3/\mu m^2$ per unit area per hour, while that for the epoxy-clay nanocomposite was nearly half at ~3.4 ×10^{-3} $\mu m^3/\mu m^2$ per unit area per hour. Also, for the durations considered, the change in material volume continues to increase. This implies that material erosion will continue to occur because of synergistic degradation.

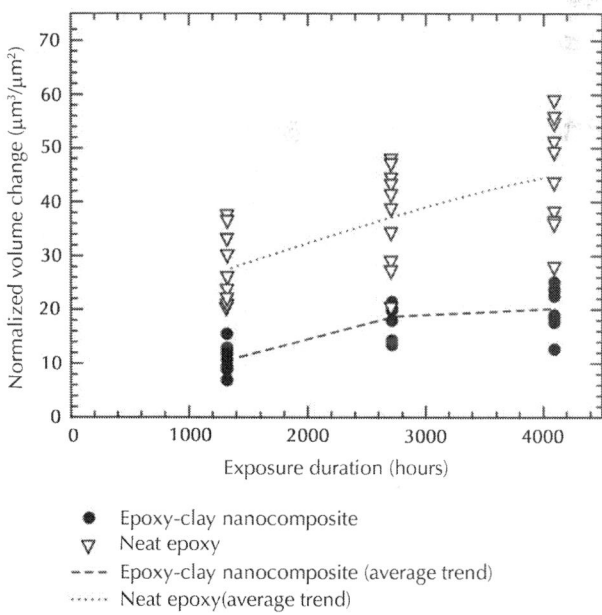

- ● Epoxy-clay nanocomposite
- ▽ Neat epoxy
- --- Epoxy-clay nanocomposite (average trend)
- ······ Neat epoxy(average trend)

Figure 13: Change in material volume, normalized with respect to the viewing area, as a function of exposure duration for neat epoxy and epoxy-clay samples exposed to combined UV radiation and condensation.

The imaging data can also be used to determine the change in material surface area. Figure 14 plots the change in material surface area, normalized with respect to the viewing area, for neat epoxy and epoxy-clay composites, as a function of exposure duration. The average rate of surface change for the neat epoxy materials was ~3.8 ×10^{-5} µm^2/µm^2 per unit area per hour, while that for the epoxy-clay nanocomposite was nearly half at ~1.5 ×10^{-5} µm^2/ µm^2 per unit area per hour. As for the material volume change, the surface change continues to increase for the durations considered.

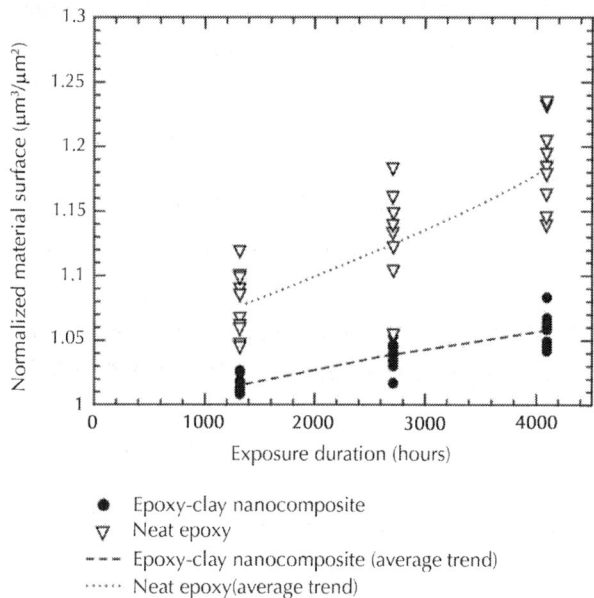

Figure 14: Change in material volume, normalized with respect to the viewing area, as a function of exposure duration for neat epoxy and epoxy-clay samples exposed to combined UV radiation and condensation.

CONCLUSIONS

The influence of nanoscale clay reinforcement of epoxy on the resistance to environmental degradation was studied. An epoxy-clay nanocomposite was fabricated using Epon 862, a bisphenol F-based epoxy resin, cured using Epikure 3274, a moderately reactive aliphatic amine, and reinforced with 4 vol.% of Nanomer I.30E, an organically modified clay. The clay particles were mixed and exfoliated with extensive mechanical mixing for 14 hours followed by high-shear dispersion for 30 minutes. Various mixing techniques were evaluated, and the selected process resulted in the greatest improvement in mechanical properties indicating complete, or near complete, exfoliation of the clay platelets.

The nanocomposites were subjected to two environmental conditions, namely, combined UV radiation and condensation on a three-hour repeat cycle and constant relative humidity. Both types of exposures were conducted for 4770 hours and at fixed temperatures of 50°C. The variation of mass loss or gain was recorded to serve as an indicator of material degradation or moisture uptake. Batches of samples were removed after 1320, 2712 and 4098 hours to test for flexure modulus and strength using four-point bend testing.

Under the exposure of constant relative humidity it was found that both neat epoxy and the epoxy-clay nanocomposite gained mass. However, the presence of nanoscale reinforcements in the epoxy matrix acted as a barrier and significantly hindered the moisture absorption. As the result, the final moisture uptake value and the rate of the moisture uptake were both measurably reduced, as expected, in comparison to neat epoxy also reduced in epoxy-clay nanocomposites.

The combined exposure of UV radiation and condensation resulted in the loss of mass of both materials due to the erosion of epoxy by a synergistic physicochemical process that was previously identified and characterized by Kumar et al. [6]. They suggested the formation of photo-oxidative byproducts that underwent dissolution by water vapor condensation and run-off results in the removal of surface layers degraded by UV radiation. Therefore, cyclic exposure to both UV radiation and water vapor condensation results in a continual material degradation and erosion process. Recently, Woo et al. also suggested that the presence of moisture can enhance the mobility of free radicals and ions and, thereby,

REFERENCES

1. G. Pritchard, Reinforced Plastics Durability, Woodhead Publishing, Cambridge, UK, 1998.

2. J. W. Chin, T. Nguyen, and K. Aouadi, "Effects of environmental exposure on Fiber-Reinforced Plastic (FRP) materials used in construction," Journal of Composites Technology and Research, vol. 19, no. 4, pp. 205–213, 1997.

3. B. Ranby and J. Rabek, Photodegradation, Photo-Oxidation and Photostabilization of Polymers, John Wiley & Sons, London, UK, 1975.

4. W. B. Liau and F. P. Tseng, "The effect of long-term ultraviolet light irradiation on polymer matrix composites," Polymer Composites, vol. 19, no. 4, pp. 440–445, 1998.

5. K.-B. Shin, C.-G. Kim, C.-S. Hong, and H.-H. Lee, "Prediction of failure thermal cycles in graphite/epoxy composite materials under simulated low earth orbit environments," Composites Part B, vol. 31, no. 3, pp. 223–235, 2000.

6. B. G. Kumar, R. P. Singh, and T. Nakamura, "Degradation of carbon fiber-reinforced epoxy composites by ultraviolet radiation and condensation," Journal of Composite Materials, vol. 36, no. 24, pp. 2713–2733, 2002.

7. A. W. Signor, M. R. VanLandingham, and J. W. Chin, "Effects of ultraviolet radiation exposure on vinyl ester resins: characterization of chemical, physical and mechanical damage," Polymer Degradation and Stability, vol. 79, no. 2, pp. 359–368, 2003.

8. C. Shen and G. S. Springer, "Moisture absorption and desorption of composite materials," Journal of Composite Materials, vol. 10, no. 1, pp. 2–20, 1976.

9. Y. Weitsman, Fatigue of Composite Materials, Elsevier, New York, NY, USA, 1991.

10. Q. Zheng and R. J. Morgan, "Synergisitc thermal-moisture damage mechanisms of epoxies and their carbon fiber

composites," Journal of Composite Materials, vol. 27, no. 15, pp. 1465–14789, 1993.

11. R. D. Adams and M. M. Singh, "The dynamic properties of fibre-reinforced polymers exposed to hot, wet conditions," Composites Science and Technology, vol. 56, no. 8, pp. 977–997, 1996. ·

12. H. S. Choi, K. J. Ahn, J.-D. Nam, and H. J. Chun, "Hygroscopic aspects of epoxy/carbon fiber composite laminates in aircraft environments," Composites Part A, vol. 32, no. 5, pp. 709–720, 2001.

13. C. Soutis and D. Turkmen, "Moisture and temperature effects of the compressive failure of CFRP unidirectional laminates," Journal of Composite Materials, vol. 31, no. 8, pp. 832–849, 1997.

14. G. Sala, "Composite degradation due to fluid absorption," Composites Part B, vol. 31, no. 5, pp. 357–373, 2000.

15. G. Mago, F. T. Fisher, and D. M. Kalyon, "Effects of multiwalled carbon nanotubes on the shear-induced crystallization behavior of poly(butylene terephthalate)," Macromolecules, vol. 41, no. 21, pp. 8103–8113, 2008.

16. E. T. Thostenson, C. Li, and T.-W. Chou, "Nanocomposites in context," Composites Science and Technology, vol. 65, no. 3-4, pp. 491–516, 2005.

17. R. A. Vaia and E. P. Giannelis, "Polymer nanocomposites: status and opportunities," MRS Bulletin, vol. 26, no. 5, pp. 394–401, 2001.

18. T. Pinnavaia, T. Lan, Z. Wang, H. Shi, and P. D. Kaviratna, Nanotechnology, vol. 622 of ACS Symposium Series, American Chemical Society, Washington, DC, USA, 1996.

19. R. Krishnamoorti and R. A. Vaia, "Polymer nanocomposites: synthesis, characterization, and modeling," in Proceedings of the 219th National Meeting of the American Chemical Society, San Francisco, Calif, USA, 2000.

20. A. Usuki, Y. Kojima, M. Kawasumi, A. Okada, T. Kurauchi, and O. Kamigaito, "Characterization and properties of nylon 6.

Clay hybrid," in Proceedings of the ACS Division of Polymer Chemistry Meeting, pp. 651–652, Washington, DC, USA, August 1990.

21. M. Alexandre and P. Dubois, "Polymer-layered silicate nanocomposites: preparation, properties and uses of a new class of materials," Materials Science and Engineering R, vol. 28, no. 1, pp. 1–63, 2000. ·

22. O. Becker, R. Varley, and G. Simon, "Morphology, thermal relaxations and mechanical properties of layered silicate nanocomposites based upon high-functionality epoxy resins," Polymer, vol. 43, no. 16, pp. 4365–4373, 2002.

23. D. Ratna, N. R. Manoj, R. Varley, R. K. Singh Raman, and G. P. Simon, "Clay-reinforced epoxy nanocomposites," Polymer International, vol. 52, no. 9, pp. 1403–1407, 2003.

24. S. C. Zunjarrao, R. Sriraman, and R. P. Singh, "Effect of processing parameters and clay volume fraction on the mechanical properties of epoxy-clay nanocomposites," Journal of Materials Science, vol. 41, no. 8, pp. 2219–2228, 2006.

25. T. Ogasawara, Y. Ishida, and T. Ishikawa, "Helium gas permeability of montmorillonite dispersed nanocomposites," in Proceedings of the 11th US-Japan Conference on Composite Materials, Yamagata, Japan, 2004.

26. Y. Kojima, A. Usuki, M. Kawasumi, A. Okada, T. Kurauchi, and O. Kamigaito, "Sorption of water in nylon 6-clay hybrid," Journal of Applied Polymer Science, vol. 49, no. 7, pp. 1259–1264, 1993.

27. T. Lan and T. J. Pinnavaia, "Clay-reinforced epoxy nanocomposites," Chemistry of Materials, vol. 6, no. 12, pp. 2216–2219, 1994.

28. T. Hwang, L. Pu, S. W. Kim, Y.-S. Oh, and J.-D. Nam, "Synthesis and barrier properties of poly(vinylidene chloride-co-acrylonitrile)/SiO2 hybrid composites by sol-gel process," Journal of Membrane Science, vol. 345, no. 1-2, pp. 90–96, 2009.

29. J.-K. Kim, C. Hu, R. S. C. Woo, and M.-L. Sham, "Moisture barrier characteristics of organoclay-epoxy nanocomposites," Composites Science and Technology, vol. 65, no. 5, pp. 805–813, 2005.

30. "Standard test method for flexural properties of unreinforced and reinforced plastics and electrical insulating materials by four-point bending," ASTM D6272–10, American Society for Testing and Materials, 2010.

31. "Standard test methods for plane-strain fracture toughness and strain energy release rate of plastic materials," ASTM D5045, American Society for Testing and Materials, 1999.

32. T. L. Anderson, "Fracture mechanics: fundamentals and applications," CRC Press 2004.

33. R. S. C. Woo, Y. Chen, H. Zhu, J. Li, J.-K. Kim, and C. K. Y. Leung, "Environmental degradation of epoxy-organoclay nanocomposites due to UV exposure. Part I: photo-degradation," Composites Science and Technology, vol. 67, no. 15-16, pp. 3448–3456, 2007.

34. R. S. C. Woo, H. Zhu, C. K. Y. Leung, and J.-K. Kim, "Environmental degradation of epoxy-organoclay nanocomposites due to UV exposure. Part II: residual mechanical properties," Composites Science and Technology, vol. 68, no. 9, pp. 2149–2155, 2008.

[2]Department of Mechanical Engineering, Islamic Azad University, Saveh Branch, Saveh, Iran

ABSTRACT

The 6061-T651 aluminium alloy is one of the most common aluminium alloys for marine components and general structures. The stress intensity factor (SIF) is an important parameter for estimating the life of the cracked structure. In this paper, the stress intensity factors of a slant-cracked plate, which is made of 6061-T651 aluminum, have been calculated using extended finite element method (XFEM) and finite element method (FEM) in ABAQUS software and the results were compared with theoretical values. Numerical values obtained from these two methods were close to the theoretical values. In simulations of crack growth at different crack angles, the crack propagation angle values were closer to the theoretical values in XFEM method. Also, the accuracy and validity of fatigue crack growth curve were much closer to the theoretical graph in XFEM than the FEM. Therefore, in this paper the capabilities of XFEM were realized in analyzing issues such as cracks.

INTRODUCTION

Fracture and failure are common problems with industry equipment. In modern materials science, fracture mechanics is an important tool in improving the mechanical performance of mechanical components. The stress intensity factor (SIF) is an important parameter for estimating the life of the cracked structure. In reality the stress intensity factor is a complicated function of applied loading, boundary conditions, crack growth, geometry, and material properties. By using the SIF and Paris law, the fatigue crack growth at the plate is measured. In fact, the Paris model describes the rate of crack growth in terms of material properties and the stress intensity factor. The stress intensity factor is performed using

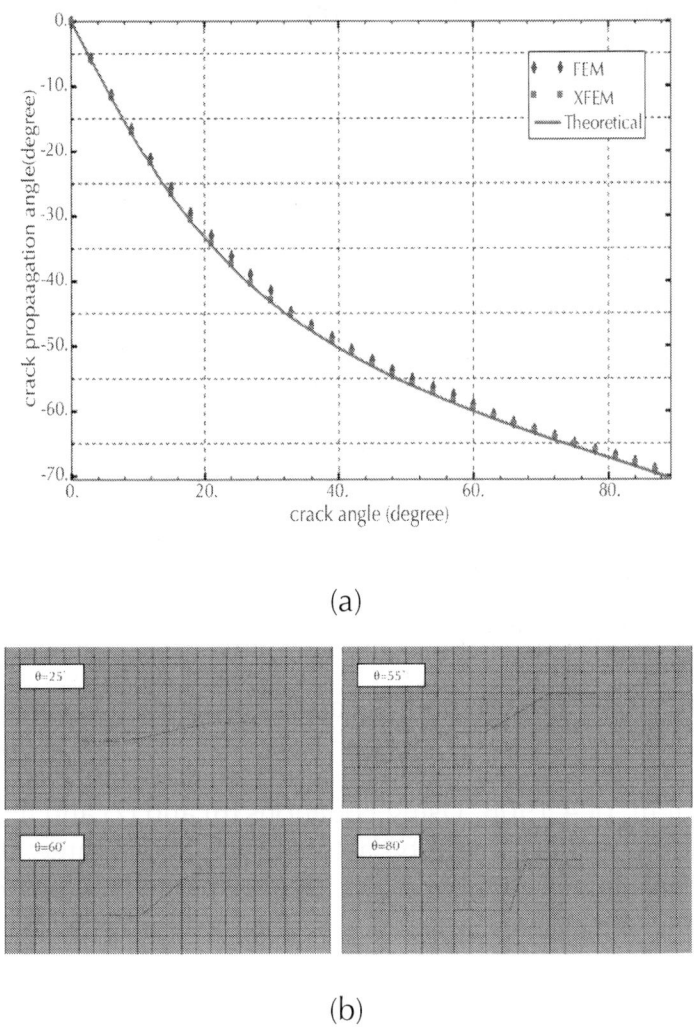

(a)

(b)

Figure 5: (a) Comparison of crack propagation angle for Different initial crack configurations; (b) Center crack propagation in the infinite plate with different initial crack configurations.

In FEM, the lifetime of 5 mm, 7 mm, and 8.8 mm cracks are 248, 1950 and 2810 cycles respectively. In XFEM, the failure values of 5 mm, 7 mm, and 8.8 mm cracks are 230, 1800 and 2600 cycles respectively. Thus, The amounts of error in FEM and XFEM are approximately 12.4 and 4 percent respectively. The results are shown in Tables 5. Also, the accuracy and validity of fatigue crack growth diagram in XFEM is closer to the theoretical method. These diagrams are presented in Figure 6 and are compared with each other. According to the overall results obtained in this paper, we can realize the capability of XFEM in the investigation of the issues such as cracks.

Table 5: Comparison of predicted Fatigue crack propagation

Crack length (mm)	θ	Theoretical N (cycles)	2D FEM N (cycles)	XFEM N (cycles)	2D FEM Error (%)	XFEM Error (%)
5	60°	221	248	230	12.217	4.072
7	60°	1730	1950	1800	12.716	4.046
8.8	60°	2500	2810	2600	12.400	4.000

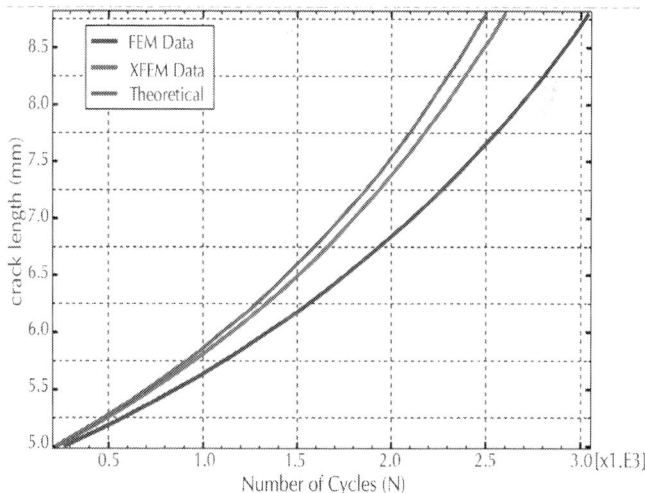

Figure 6: Theoretical, 2D FEM and XFEM crack growth curves.

7. M. Stolarska, D. L. Chopp, N. Moës and T. Belytschko, "Modelling Crack Growth by Level Sets in the Extended Finite Element Method," International Journal for Numerical Methods in Engineering, Vol. 51, No. 8, 2001, pp. 943-960. http://dx.doi.org/10.1002/nme.201

8. N. Sukumar and J. H. Prevost, "Modeling Quasi-Static Crack Growth with the Extended Finite Element Method Part I: Computer Implementation," International Journal of Solids Structure, Vol. 40, No. 26, 2003, pp. 7513-7537.http://dx.doi.org/10.1016/j.ijsolstr.2003.08.002

9. N. Sukumar, R. Huang and J. H. Prevost, "Modeling Quasi-Static Crack Growth with the Extended Finite Element Method Part II: Numerical Applications," International Journal of Solids Structure, Vol. 40, No. 26, 2003, pp. 7539-7552.http://dx.doi.org/10.1016/j.ijsolstr.2003.08.001

10. N. Sukumar, D. Baker. T. Srolovitz and J. Prevost, "Brittle Fracture in Polycrystalline Microstructures with the Extended Finite Element Method," International Journal for Numerical Methods in Engineering, Vol. 56, No. 14, 2003, pp. 2015-2037.http://dx.doi.org/10.1002/nme.653

11. C. Daux, N. Moes, J. Dolbow and N. Sukumar, "Arbitrary Branched and Intersecting Cracks with the Extended Finite Element Method," International Journal for Numerical Methods in Engineering, Vol. 48, No. 12, 2000, pp. 1741-1760.http://dx.doi.org/10.1002/1097-0207(20000830)48:12<1741::AID-NME956>3.0.CO;2-L

12. N. Sukumar, N. Moes, B. Moran and T. Belytschko, "Extended Finite Element Method for Three-Dimensional Crack Modeling," International Journal for Numerical Methods in Engineering, Vol. 48, No. 11, 2000, pp. 1549- 1570. http://dx.doi.org/10.1002/1097-0207(20000820)48:11<1549::AID-NME955>3.0.CO;2-A

13. Abaqus 6.10, Analysis User's Manual Volume Number 2: Analysis, Dassault simulia.

14. C. Zhang, P. Cao, Y. Cao and J. Li, "Using Finite Element Software to Simulation Fracture Behavior of Threepoint

Bending Beam with Initial Crack," Journal of software, Vol. 8, No. 5, 2013, pp. 1145-1150. http://dx.doi.org/10.4304/jsw.8.5.1145-1150

15. A. Sutradhar and G. H. Paulino, "Symmetric Galerkin Boundary Element Computation of T-Stress and Stress Intensity Factors for Mixed-Mode Cracks by the Interaction Integral Method," Engineering Analysis with Boundary Elements, Vol. 28, No. 11, 2004, pp. 1335-1350. http://dx.doi.org/10.1016/j.enganabound.2004.02.009

16. T. Belytschko and T. Black, "Elastic Crack Growth in Finite Elements with Minimal Remeshing," International Journal for Numerical Methods in Engineering, Vol. 45, No. 5, 1999, pp. 601-620. http://dx.doi.org/10.1002/(SICI)1097-0207(19990620)45:5<601::AID-NME598>3.0.CO;2-S

17. A. S. Ribeiro and M. P. de Jesus Abílio, "Fatigue Behavior of Welded Joints Made of 6061-T651 Aluminum Alloy," In: T. Kvackaj, Ed., Aluminum Alloys, Theory and Applications, InTech, 2011. http://www.intechopen.com/books/aluminium-alloys-theory-and-applications/fatigue-behaviour-of-welded-joints-made-of-6061-t651-aluminium-alloyhttp://dx.doi.org/10.5772/14489

18. J. Yau, S. Wang and H. Corten, "A Mixed Mode Crack analysis of Isotropic Solids Using Conservation Laws of Elasticity," Journal of Applied Mechanics, Vol. 47, No. 2, 1980, pp. 335-341. http://dx.doi.org/10.1115/1.3153665

19. E. Giner, N. Sukumar, J. E. Tarancon and F. J. Fuenmayor, "An Abaqus Implementation of the Extended Finite Element Method," Fracture Mechanic, Vol. 76, No. 3, 2009, pp. 347-368. http://dx.doi.org/10.1016/j.engfracmech.2008.10.015

20. F. Erdogan and G. Sih, "On the Crack Extension in Plates under Plane Loading and Transverse Shear," ASME Journal of Basic Engineering, Vol. 85, No. 4, 1963, pp. 519- 527. http://dx.doi.org/10.1115/1.3656899

21. P. C. Paris and F. Erdogan, "A Critical Analysis of Crack Propagation Laws," ASME Journal of Basic Engineering,

Vol. 85, No. 4, 1963, pp. 528-527.http://dx.doi. org/10.1115/1.3656900

22. K. Tanaka, "Fatigue Crack Propagation from a Crack Inclined to the Cyclic Tensile Axis," Engineering Fracture Mechanics, Vol. 6, No. 3, 1974, pp. 493-507.http://dx.doi.org/10.1016/0013-7944(74)90007-1

Estimation of Fatigue Life of Laser Welded AISI304 Stainless Steel T-Joint Based on Experiments and Recommendations in Design Codes

Søren Heide Lambertsen, Lars Damkilde, Anders Schmidt Kristensen, and Ronnie Refstrup Pedersen

Division of Structures and Materials, Aalborg University Esbjerg, Esbjerg, Denmark

ABSTRACT

In this paper the fatigue behavior of laser welded T-joints of stainless steel AISI304 is investigated experimentally. In the fatigue experiments 36 specimens with a sheet thickness of 1 mm are exposed to one-dimensional cyclic loading. Three different types of specimens are adopted. Three groups of specimens are used, two of these are non-welded and the third is welded with a transverse welding (T-Joint). The 13 laser welded specimens are cut out with a milling cutter. The non-welded specimens are divided in 13 specimens cut out with a milling cutter and 10 specimens cut out by a plasma cutter. The non-welded specimens are used to study the influence of heat and surface effects on the fatigue life. The fatigue life from the experiments is compared to fatigue life calculated from the guidelines in the standards DNV-RP-C203 and EUROCODE 3 EN-1993-1-9. Insignificant differences in fatigue life of the welded and non-welded specimens are observed in the experiments and the largest difference is found in the High Cycle Fatigue (HCF) area. The specimens show a lower fatigue life compared to DNV-RP-C203 and EUROCODE 3 EN-1993-1-9 when the specimens are exposed to less than 4.0 1E06 cycles. Therefore, we conclude that the fatigue life assessment according to the mentioned standards is not satisfactory and reliable.

INTRODUCTION

S-N curves are based on experimental data, where traditional welding methods are adopted. Therefore, the S-N curves may not correctly represent the fatigue life, where the laser welding method and other non-traditional welding methods are applied. Nowadays the high-speed laser weld method is widely used in the industry. Hence the standards might need new S-N curves that correctly represent the fatigue life of laser welded stainless steel materials. Laser welding is commonly used to assemble small components in the biomedical, electronics and aerospace industry [1].

In these applications weldings require a very small melted area, hence small laser beams less than 1 kW are used. The Nd:YAG laser beam has been developed further and the maximum heat output is increased to 6 kW [2]. Consequently, it is possible to weld sheet components with a higher thickness. Today AISI304 sheets with a thickness up to 12 mm can be assembled with a modern Nd:YAG laser. Therefore, the application area of the laser welding method is expanded to include traditional welding tasks and the laser welding replaces the TIG and MIG/MAG welding methods.

One of the most significant types of laser welding is the keyhole method [3], which is also used in this study. With the keyhole method it is possible to weld, in one process, perpendicular plates. Therefore, in production of perpendicular plate assemblies the keyhole welding method permits high production volume and a low manufacturing price. The possibility to weld a perpendicular assembly in one single process has been requested by the industry for many years. Therefore, the keyhole method became one of the most important welding procedures with the laser technology. However, when the keyhole welding is used higher stress concentrations are generated in the welded zone and consequently a lower fatigue life can occur. The fatigue behavior of dynamic loaded structures, where the keyhole method is adopted has not been investigated thoroughly in the literature. Therefore, it is necessary to investigate the fatigue life of T-joints exposed to dynamic loading. Laser welding is more effective compared to traditional welding methods, i.e., the process is faster and more energy in supplied to a smaller welding area [4]. However, the small melted area cools down rapidly and high thermal stresses are introduced [5]. The thermal stress influences the metallurgy in the welding area and thereby the mechanical properties [6].

Guidelines

The present guidelines for fatigue assessment of different types of welded joints according to the design standards are quite general. For stainless steel the assessment is independent of several important aspects mentioned next. Geometrically, the laser welding also cause a smaller welding compared to traditional welding types. However, the smaller welding toe implies in most cases higher stress concentrations and accordingly the fatigue life is decreased. Furthermore, rotating bending fatigue tests [7] with

AISI304 materials have shown that the fatigue life is sensitive to the initial defects in the material. The higher stress concentrations are not properly accounted for in the current design guidelines because they are based on experiments with relatively smooth welding's. The guidelines suggest that an averaged thickness of the sheets and an averaged size of welding toes are used in the fatigue assessment. A fatigue calculation based on these averaged values may not provide a reliable estimate of the fatigue life. The size of the welding toe is known to have a greater impact on the fatigue life in a thin sheet application compared to a situation, where a thick sheet is used [6]. The higher sensitivity in thin plates is caused by the stress concentration in the normal plane to the specimen. When the welding toe is large compared to the plate thickness the toe has a considerable influence on the fatigue life [8] and [9]. A reliable estimate of the fatigue life requires that the geometry of the welding toe is taken into account in the design guidelines.

Material

In the design standards the fatigue assessment guidelines are based on structural steel like S235JR. These types of steel do not show the same tendency in change of the microstructure under the maximum crack length is decreased. However, the crack growth rate is also decreased. The induced strains ahead of a fatigue crack tip can activate a transformation of an austenitic structure to welding process compared to stainless steel. Therefore, we do not expect that the fatigue life calculated based on the standards fit the fatigue life observed in the experiments, where thin sheets of stainless steel AISI304 are tested. In some types of stainless steel including AISI304 the microstructure is changed i.e. austenitic phase is transformed into a martensitic phase. This transformation is mainly controlled by plastic strain rates and the temperature [10-12]. In case of low temperatures a higher amount of martensitic phase is generated [13-15]. The martensitic phase is more brittle than the austenitic phase. Therefore, the fracture toughness limit is lower and a martensitic structure [16-20]. Furthermore, this transformation also changes the volume and the expansion of the material can lead to compressive stresses. Hence deformation induced martensitic transformation increases fatigue resistance significantly and the threshold stress intensity decreases. The standard S-N curves for welded structural steel do not include the mean stress because of welding introduced residual stresses. In stainless steel the residual stresses coming from the laser welding process are often more significant compared to standard structural steel. Therefore, the higher compressive stresses in the welding zone can lead to crack propagation resistance [21-23]. However, the tensile stresses in the welding zone are increased accordingly and crack initiations are more probable.

The Study

In this study the fatigue life of welded and non-welded specimens is investigated experimentally. The welded specimens are produced with the keyhole laser welding method. The results from the experiments are compared to the results from the standards, where the S-N curves are based on structural steel and traditional welding methods. Hence this comparison and possible deviations will reveal if the guidelines for fatigue assessments in the current standards can be safely adopted when stainless steel is considered.

EXPERIMENTAL DETAILS

The stainless steel used in the experiments is AISI304 and the sheets have a thickness of 1 mm. The welded specimen is a T-joint welded with the keyhole method by a laser and is shown in Figure 1. The concentration of the elements Cr, Ni, S, Mn, Mo, Si and C of the austenitic phase are determined by Electron Probe Micro Analysis. The AISI304 composition is 0.036% C, 0.42% Si, 1.28% Mn, 0.031% P, 0.0010% S, 18.21% Cr, and 8.30% Ni.

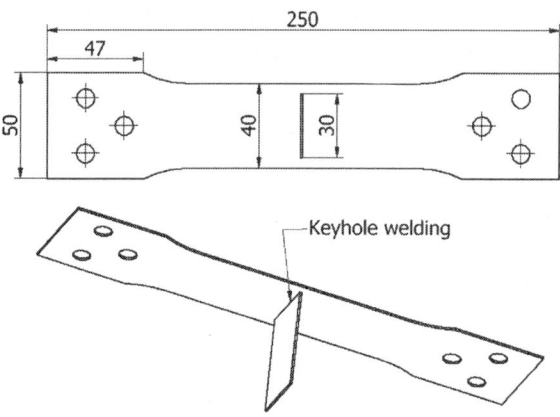

Figure 1: The geometry of the laser welded specimen in AISI 304.

The mechanical properties are $\sigma_{p0.2\%}$ = 279 MPa, $\sigma_{p1\%}$ = 310 MPa, σ_U = 635 MPa ultimate strength and 57% elongation. Three different types of specimen are used in the fatigue tests. Two types are non-welded, where one type is cut with a plasma cutter, whereas the other nonwelded type is cut with a milling cutter. The third type is a specimen with a transverse welding with respect to the load direction (T-joint). The welding is carried out with a 2 kW NdYAG laser with a speed of 40 mm per second. Argon gas is used to prevent oxidation of the steel. The steel surfaces are smooth with a maximum variation of 0.01 mm. The welded surfaces have a variation around 0.1 mm. A standard servohydraulic Instron 1255 test machine is used. The experiments are load controlled with a constant amplitude sinusoidal wave form with R = 0.1 and load frequencies in the interval (2 - 12) Hz. The test is completed when a final fracture occurs. The temperature in the laboratory is in the interval (19°C - 23°C). The geometry is hourglass formed with a design approximation of ASTM E466 and E468. Two different methods have been used to cut out the non-welded specimens. The different defects from this cutting process are observed and the influence on the fatigue life is established. The knowledge about these defects can be transformed to a tolerable defect size for the S-N curve.

In Figure 2 the non-welded specimen geometry cut by the milling cutter is shown and Figure 3 shows the geometry cut by the plasma cutter.

The missing data points do to the limit number of sample is estimated with interpolation between the low and high cycle fatigue life.

RESULTS

The shape of the welding significantly influences stress concentrations. Thus the shape of the welding is an important factor in the estimation of the fatigue life. A macro photo of the cross section of the welding is shown in Figure 4. The welding has clear defects, which introduce high stress concentrations. The shape is like a wedge in the area where the two parts are welded together. A crack most probably initiates at the end of the wedge shape and the fatigue life will be lower compared to a specimen with a smooth welding. No materials are added during the welding process, which explains these clear defects shown in Figure 4.

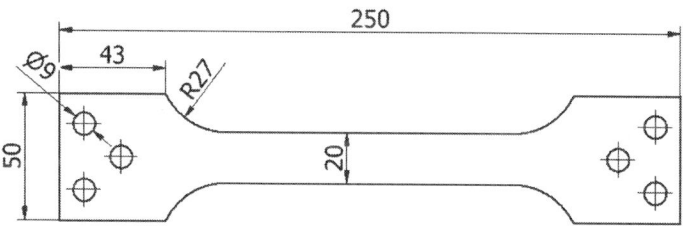

Figure 2: The geometry of the modified non-welded specimens cut out with a milling cutter.

Clearly the difference in fatigue life is small. In Figure 9 the fatigue life of the non-welded specimens is compared with the fatigue life calculated on basis of the guidelines in EUROCODE 3 EN-1993-1-9 and DNV RP-C203.

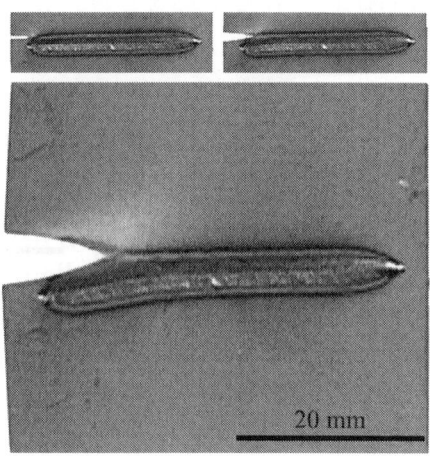

Figure 6: The edge crack growth at the laser welded specimen. The cracks tend to grow in the HAZ Zone.

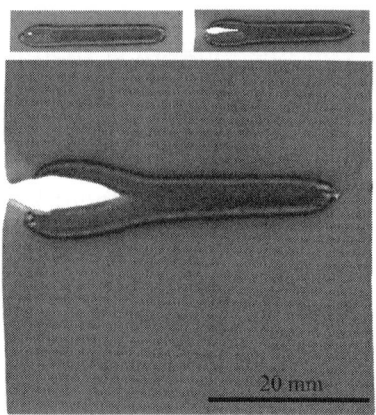

Figure 7: The center crack in the welding at the laser welded specimen. The crack grows in the pre-crack area between the two parts.

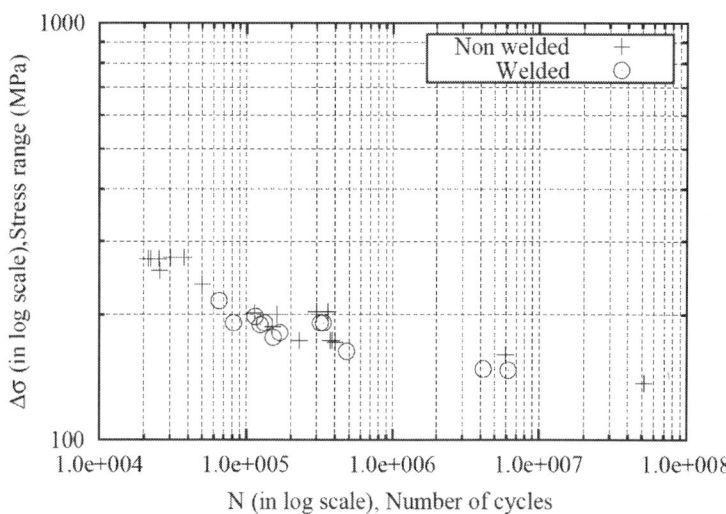

Figure 8: Plot of the fatigue life for the welded and nonwelded specimens.

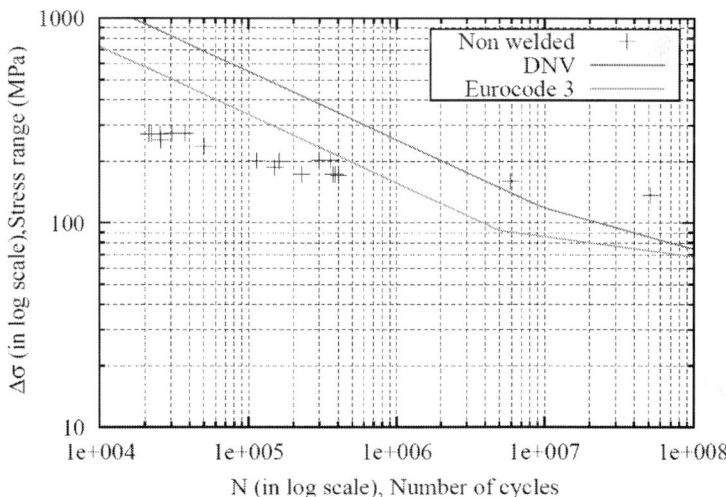

Figure 9: Plot of the fatigue life for the non-welded specimens and the estimated fatigue life based on EUROCODE 3 EN-1993-1-9 and DNV-RP-C203.

The curves in the DNV RP-C203 standard are associated with a 97.7% probability of survival, whereas the curves in the EUROCODE 3 EN-1993-1-9 standard are associated with a 95% probability of survival. Therefore, the test data should provide higher fatigue strength to compensate for the safety incorporated in the standards. In the calculations according to EUROCODE 3 EN-1993- 1-9 (see Equation (1)) the detail category 125 is used, where m = 3 below constant amplitude limit at 5.0 1E06 cycles and m = 5 up to 5.0 1E08 which is the cut-off limit.

$$\Delta\sigma_R^m N_R = \Delta\sigma_c^m 2\times10^6$$

(1)

The calculations with the DNV RP-C203 standard (see Equation (2)) are based on design curve C, where m = 3 and log (a) = 12.592 below 1E07 cycles and m = 5 and log (a) = 16.320 above 1E07 cycles. The thickness exponent k = 0.15 with a reference thickness of 25 mm.

$$\log N = \log a - m\log\left(\Delta\sigma\left(t/t_{ref}\right)^k\right)$$

(2)

In Figure 10 the curve from EUROCODE 3 EN-1993- 1-9 is designed from detail category 80, where m = 3 below the constant amplitude limit at 5.0 1E06 cycles and m = 5 up to 1E08 cycles, which is the cut-off limit. The DNV curve is based on design curve E, where m = 3 and log (a) = 12.01 below 1E07 cycles and m = 5 and log (a) = 15.35 above 1E07 cycles. The thickness exponent k = 0.20 with a reference thickness of 25 mm and a structural stress concentration factor of 1.13.

It is clear from Figures 9 and 10 that the calculated fatigue life gives a long estimate life at the HCF area. With lower cycles fatigue the fatigue life for the specimens tend to have less resistance to a dynamic load. The negligible difference in fatigue life for the welded and nonwelded specimens is most probably caused by local relaxation of the material. The residual stresses from the welding process relax the material by local compression areas and thereby extend the fatigue life. Microstructural transformations can also change the fatigue life. It has been observed that AISI304 can change microstructure from austenite to martensitic and thereby change fatigue and fracture parameters like K_{IC}, K_C and crack growth speed. The residual stress and the microstructure transformation are most probable the source that influence the fatigue life so it is similar for welded and non-welded specimens.

The slope of experimental data and of the design code is not equal. The design code is based on ferritic steel and the material used in the experiments is austenitic stainless steel. In the Stainless steel a considerable amount of chrome and nickel affect the fracture parameter and microstructure. This is the reason for the variation of the slope on the S-N curve.

CONCLUSIONS

In this paper the experimental investigations of the fatigue life for laser welded T-joints and non-welded specimens made of AISI304 stainless steel are presented. Based on these experimental results the main conclusions are:

- The difference in the fatigue life of a specimen generated by plasma cut or milling cut is insignificant.
- The largest difference in the fatigue life for the plasma and milling cut shows to be in the high cycle fatigue area.
- A comparison between the fatigue life of the welded T-joints specimens and the non-welded specimens shows the tendency that the fatigue resistance is higher for the non-welded specimens.
- The results show that the fatigue life for the welded specimens is lower than the fatigue life estimation based on the standards.

Processing Technology, Vol. 209, No. 8, 2009, pp. 4011-4019.

10. X. Yang, J. Zhou and X. Ling, "Study on Plastic Damage of AISI304 Stainless Steel Induced by Ultrasonic Impact Treatment," Materials and Design, Vol. 36, 2012, pp. 477-481. doi:10.1016/j.matdes.2011.11.023

11. M. C. Park, K. N. Kim, G. S. Shin and S. J. Kim, "Effects of Strain Induced Martensitic Transformation on the Cavitation Erosion Resistance and Incubation Time of Fe-CrNi-C Alloys," Wear, Vol. 274-275, 2012, pp. 28-33. doi:10.1016/j.wear.2011.08.011

12. M. Jayaprakash, J. Sumanth Kumar, S. Katakam and S. G. S. Raman, "Effect of Grain Size on Fretting Fatigue Behaviour of Aisi 304 Stainless Steel," International Symposium of Research Students on Materials Science and Engineering, Chennai, 20-22 December 2004, pp. 1-8. http://mme.iitm.ac.in/isrs/isrs04/cd/content/Papers/MBM/PO-MBM-8.pdf

13. N. Rossinia, M. Dassistia, K. Benyounisb and A. Olabib, "Methods of Measuring Residual Stresses in Components," Materials and Design, Vol. 35, 2012, pp. 572-588. doi:10.1016/j.matdes.2011.08.022

14. C. Müller-Bollenhagen, M. Zimmermann and H.-J. Christ, "Very High Cycle Fatigue Behavior of Austenitic Stainless Steel and the Effect of Strain-Induced Marten-Site," International Journal of Fatigue, Vol. 32, No. 6, 2010, pp. 936-942. doi:10.1016/j.ijfatigue.2009.05.007

15. O. Takakuwaa, M. Nishikawab and H. Soyama, "Numerical Simulation of the Effects of Residual Stress on the Concentration of Hydrogen around a Crack Tip," Surface and Coatings Technology, Vol. 206, No. 11-12, 2012, pp. 2892-2898. doi:10.1016/j.surfcoat.2011.12.018

16. O. Keiji, M. Yoshio and N. Izuru, "Threshold Behavior of Small Fatigue Crack at Notch Root in Type AISI 304 Stainless Steel," Engineering Fracture Mechanics, Vol. 25, No. 1, 1986, pp. 31-46. doi:10.1016/0013-7944(86)90201-8

17. M. C. Young, J. Y. Huang and R. C. Kuo, "Corrosion Fatigue Behavior of Cold-Worked 304L Stainless Steel," Materials Transactions, Vol. 50, No. 3, 2009, pp. 657- 663.

18. M. C. Park, K. N. Kim, G. S. Shin and S. J. Kim, "Effects of Strain Induced Martensitic Transformation on the Cavitation Erosion Resistance and Incubation Time of FeCr-Ni-C Alloys," Wear, Vol. 274-275, 2012, pp. 28-33. doi:10.1016/j.wear.2011.08.011

19. L. Singh, R. A. Khan and M. L. Aggarwal, "Influence of Residual Stress on Fatigue Design of AISI 304 Stainless Steel," The Journal of Engineering Research, Vol. 8, No. 1, 2011, pp. 44-52.

20. C. Mu□ller-Bollenhagen, M. Zimmermann and H.-J. Christ, "Adjusting the Very High Cycle Fatigue Properties of a Metastable Austenitic Stainless Steel by Means of the Martensite Content," Procedia Engineering, Vol. 2, No. 1, 2010, pp. 1663-1672.

21. L. Tsay, Y. Liu, D. Y. Lin and M. Young, "The Use of Laser Surface-Annealed Treatment to Retard Fatigue Crack Growth of Austenitic Stainless Steel," Materials Science and Engineering, Vol. 384, No. 1-2, 2004, pp. 177-183. doi:10.1016/j.msea.2004.06.010

22. A. Hascalik, E. Unal and N. Ozdemir, "Fatigue Behaviour of AISI 304 Steel to AISI 4340 Steel Welded by Friction Welding," Journal of Materials Science, Vol. 41, No. 11, 2006, pp. 3233-3239. doi:10.1007/s10853-005-5478-7

23. Y. C. Chiou, "Experimantal Study of Deformation Behavior and Fatigue Life of AISI304 Stainless Steel under an Asymmetric Cyclic Loading," Journal of Marine Science and Technology, Vol. 18, No. 1, 2010, pp. 122-129.

6

Hybrid Carbon-Carbon Ablative Composites for Thermal Protection in Aerospace

P. Sanoj and Balasubramanian Kandasubramanian

Department of Materials Engineering, Defence Institute of Advanced Technology (DIAT) (DU), Girinagar, Pune, Maharashtra 411025, India

ABSTRACT

Composite materials have been steadily substituting metals and alloys due to their better thermomechanical properties. The successful application of composite materials for high temperature zones in aerospace applications has resulted in extensive exploration of cost effective ablative materials. High temperature heat shielding to body, be it external or internal, has become essential in the space vehicles. The heat shielding primarily protects the substrate material from external kinetic heating and the internal insulation protects the subsystems and helps to keep coefficient of thermal expansion low. The external temperature due to kinetic heating may increase to about maximum of 500°C for hypersonic reentry space vehicles while the combustion chamber temperatures in case of rocket and missile engines range between 2000°C and 3000°C. Composite materials of which carbon-carbon composites or the carbon allotropes are the most preferred material for heat shielding applications due to their exceptional chemical and thermal resistance.

INTRODUCTION

Discovery of carbon-carbon composites in 1958 by Brennan Chance Vought Aircraft created an opportunity to these principle materials for heat shielding appliances due to their high strength and thermal resistance [1]. Rayon carbon fabric reinforced phenolic (C–Ph) composites are the broadly used thermal protection systems due to the low thermal conductivity of the rayon fabric and high char yields of the phenolic resin. In general, carbon phenolic composites show better ablation resistance and continued enhancement of ablative property with the development of a thinner ablative composite structure for better pay load and fuel efficiency [2]. The Space Shuttle Columbia disaster occurred on February 1, 2003, due to the inadequate impact resistance of the thermal insulation foam in the external tank against air, as the spacecraft reentered the earth's planetary atmospheric domain. The displaced reinforcement foam damaged Columbia's left reinforced carbon-carbon (RCC) panels thereby causing the unfortunate accident. This incident paves way for a detailed research to enhance impact tolerances, thermal resistance, and fracture toughness of the RCC panels [3]. Polymer nanocomposites are the three phase composite systems invented by Toyota research group, wherein nanosize particles, dispersed in the two phase fiber reinforced composites, exhibit enhanced structural rigidity and ablation resistance [1]. Nanocomposites have the capability to withstand the simultaneous action of thermal stresses and mechanical impact loads. Addition of various nanoparticles such as nanosilica, montmorillonite (MMT), nanoclays, and polyhedral oligomeric silsesquioxane (POSS) with surface functionalization acts as thermal insulative elements for improving char layer integrity and toughness. Three phase composite system with heterogeneous composition (fiber reinforcement, matrix, and nanofillers) exhibits complexity in its ablation behaviour [4]. The scientific insights into the ablation and decomposition behaviour of the composite materials lead us to a trustworthy analysis of the composite performance at high temperature working environment. This review has focused on the recent developments in the carbon-carbon composites and resultant thermal protection mechanisms. Microstructural changes during the transition from two phase composite system to three phases have been discussed in detailed manner.

STRUCTURE DESCRIPTION OF HOT ZONE ASSEMBLY

Missile structures are the extremely crucial components in aerospace industry. They should have high structural integrity and fire resistance against severe lateral pressure and thermal cyclic loading. Nozzle, the exhaust duct of the solid rocket motor, gives thrust for the missile projectile motion [5]. During missile flight, nozzle experiences an impact of jet flow with high temperature and pressure. Since even slight degradation in nozzle structure severely affects the engine performance, structural integrity is the main concern during nozzle operational period. The nozzle liner is the lining system in the nozzle inner counter to form insulation barrier. The high temperature fluid flow in the inner contour of the nozzle system creates a soaring thermomechanical stresses through the liner cross section. This degrades the nozzle throat and widens the throat area of cross section [6]. This phenomenon causes a reduction in thrust and nozzle operational efficiency. Erosion rate of the composite system depends on the thermal insulation capacity of the ablative composite char layer. Minimum surface recession rate of the char layer shows an efficient thermal shielding capacity of the ablator. Windhorst and Blount have revealed the usage of pyrolytic graphite as thermal insulation and its brittle failure due to the thermomechanical stresses [7]. It is reported that phenolic impregnated carbon composite is a popular ablative material for rocket nozzle liners due to its shielding against combustion flame and high velocity erosive fluid flow [8]. These extreme working environments demand high structural integrity for nozzle liner with minimum structural degradation.

ABLATIVE MATERIALS

Figure 1 shows the constituent elements in the carbon-carbon composites.

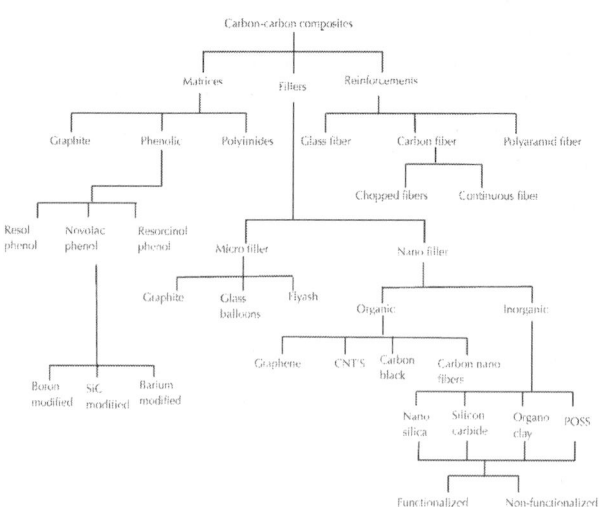

Figure 1: Carbon-carbon composite system and its constituent elements [8].

Carbon Fibres

Carbon fiber is a synthetic fiber reinforcement with a diameter of about 5–10µ m. The crystalline arrangement of the carbon atoms parallel to the fiber axis enables high strength-to-volume ratio of the fiber with excellent applications for structural components [9]. The precursors used for the large-scale production of carbon fibers are PAN, Rayon, and petroleum pitch through melt or solution spinning. PAN-based fibres show better mechanical strength as compared to the other fibers [10]. The carbon felt has been fabricated by alternatively stacked weftless piles and short-cut-fiber webs by needle-punching technique. This technique minimizes the fiber bend and breakage with the capability to tailor the properties by weaving in appropriate directions. Normally nonwoven carbon composite shows highest tensile strength over woven materials [11]. Weave pattern orientation of the carbon fabric affects the heat diffusion rate through the cross section of the composite. Kuo and Keswani studied the changes in the weave pattern of carbon fibers in different directions and altered specific properties of carbon composite [5].

Surface-treatment of the carbon fibres is a current advancement in the fiber technology, in which surface treatment by means of physical or chemical method improves the adhesion between carbon fiber and polymer matrix [12]. Production of carbon fiber through fiber spinning is highly expensive process. The usage of mesophase pitches shows efficient fiber reinforcement with improved cost effectiveness. Chand has described the mesophase pitches, which are of liquid crystalline nature. During graphitization step, mesophase pitches form a graphitic crystalline structure with high modulus carbon fibers with high stiffness [11]. Strength improvement in the fibre reinforcement is mandatory for ablative composites due to dependency of the ablation rate of composite with the fiber reinforcement morphology. More research needs to be focused over the enhancement in fiber morphology, design of fiber laminate orientation, and cost effectiveness in fiber production.

Vapour Grown Carbon Nanofiber

Vapour grown carbon fibers (VGCFs) are the recent desirable materials in thermal shielding composites due to their superior thermal insulation, low thermal conductivity (0.45–0.58 W/mk), and high thermal shock resistance [13]. These nanofibers (VGCNFs) are cylindrical nanostructures with graphene layers arranged in a stacked cones shape. Synthesis of the VGCNFs is carried out through decomposing gas-phase molecules at high temperature (deposition of carbon in the presence of a transition metal catalyst on a substrate) [14]. Tibbetts et al. grew a series of different diameter (7–30 µm) vapour grown carbon fibers and from these carbon fibers it was observed that young's modulus and tensile strength decreased with increase in diameter of fibers [15]. Dispersion and homogeneity are the specialty of VGCNF in the composite system. It leads to uniformity of properties throughout the volume of composite. Figures 2(a) and 2(b) illustrate the fiber morphology of the vapour grown carbon fibers in different magnification. Patton et al. confirmed that 65% vapour grown carbon fiber loading in VGCF composite get a good thermochemical ablation resistance for space shuttle reusable solid rocket motor [13]. Recent research on the effect of different oxidative surface treatments (nitric acid, plasma, air, and carbon dioxide) on the fibres surface shows the effectiveness of the nitric acid and plasma treatments for improving the surface reactivity without altering the morphology of the fibres. This enhances the adhesion of VGCNFs to the phenolic matrix system [16]. Fiber orientation in Vapour Grown Carbon fiber is a critical factor during the modelling of the composite system [17, 18]. This can be statistically modelled by an orientation distribution function describing the probability of finding fibres with any given orientation [19]. Advani and Tucker introduced a compact tensor description for fibre orientation, which allows easy integration with conventional rheological and mechanical tensor descriptions. Due to these advantages, it is now widely used in works on short fibre composites [20].

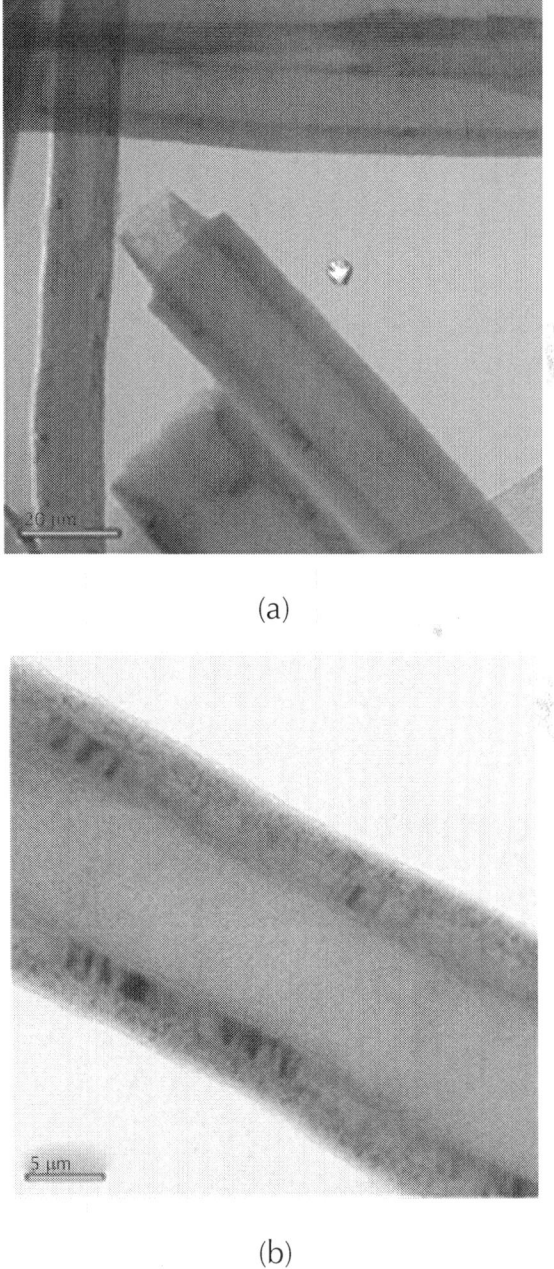

(a)

(b)

Figure 2: TEM image of Vapour Grown Carbon Fibers (a) magnification 20 nm, (b) magnification 5 nm [21].

Carbonaceous Matrix

Phenolic Resin

Matrix materials in the carbon-carbon composite have significant functional properties as it holds the reinforcement fiber and impart structural integrity to the composite system. Resol type Phenolic systems have attracted great scientific interest as they can be effectively used as a matrix system in ablation resistant composites [21]. During the fire exposure, phenolic resin receives heat in the initial stage and decomposes to char and forms a thermal insulation layer. Large number of aromatic rings in the Resol type phenolic resin results high carbon yield and effective char formation ability [27]. High char retention of the Resol type phenolic resins makes them an effective applicant for ablative nozzle liner application. Presence of hydroxyl and methyl linkages in the phenolic is prone to oxidation and demands further modification in the phenolic resin compound for practical usage as thermal ablators [28]. Chemical modifications have been well investigated to improve oxidation resistance of phenolic resin. Modification of phenolic resin with elements such as Boron, Titanium, Molybdenum, and Phosphorous shows better ablation resistance and char yield [29]. Improvement in the tribological properties of the phenolic resin enhances the wear and aerodynamic shear resistance of the phenol based carbon composites [30]. Yi and Yan studied the mechanical and tribological properties of phenolic composite dispersed with several inorganic fillers like calcined petroleum coke (CPC), talcum powder (TP), and hexagonal boron nitride (h-BN) [31]. The phenol based composite with 10% h-BN shows excellent friction stability and wear resistance at various testing conditions beyond 125°C and results in formation of compact friction film on the rubbing surface of composite. There is always a constant effort for improving the processability, toughness, and char yield of the phenolic resin by compounding with highly stable additives such as SiC, Boron nitride, Nanosilica, and zirconium diboride [32]. This modification led to better clasping of the turbostratic carbon, formed as a result of high temperature pyrolysis of hydrocarbons, consequently reducing the erosive loss of the char layer [33]. In recent years, Polyhedral Oligomeric Silsesquioxane finds attention due to their nanosize in organic cage like cluster morphology which produce best charred surface on burnt samples with an enhanced ablation performance in carbon-carbon composite [34].

Phenolic Based Carbon Foam

Phenolic foams are the light weight ablators with an exceptional fire resistance used in thermal insulation system for aerospace application. Cost effectiveness and absence of dripping molten plastic at the stage of combustion are the added advantages of phenolic based carbon foams. Generally, pure phenolic resins foam does not have the required strength to withstand the aerodynamic stresses and thermal fluctuations due to its friability [35]. Phenolic foam reinforced with synthetic fibers is the recent potential material for thermal insulator applications to improve composite strength, toughness, and effective fire resistance. Zhou et al. developed lightweight glass fiber reinforced phenolic foam with high mechanical performance and excellent flame resistance and noticed the improvement in storage modulus of the foam with reduction in loss modulus [36].

The current research is primarily focusing on the improvement in toughness and peel strength of the carbon phenolic foams through reinforcement of flexible synthetic fibers like Glass and Kevlar. Short chopped glass fibers treated with coupling agents reportedly increased the strength, toughness, and dimensional stability of phenolic foams. However, Shen et al. studied the improvements in peel strength of carbon phenolic foams with Kevlar fiber reinforcement and obtained multiple-fold enhancement in fracture toughness [37].

Various research groups have attempted to modify the phenolic foam through nano- and microsize filler dispersion like carbon black, talc, nanosilica, fly ash, asbestos, and cork flours. Exfoliated phenolic resin/montmorillonite nanocomposite shows excellent thermal insulation behaviour as Montmorillonite (MMT) layers wrap around the bubbles of the composite foams during high temperature environment. John et al. developed medium-density foam composites based on silica fibre-filled phenolic resin which shows maximum tensile strength at silica fibre concentration of 15% by volume with K37 glass balloons [22]. The reinforcement of silica fibre phenolic foam with microglass balloons before and after the aerothermal test (the black colour on the surface of the sample is an emissivity coating) is illustrated in Figures 3(a) and 3(b), respectively. Nucleating agents are the effective materials used to reduce bubble size with enhancement of strength and toughness of phenolic foam. Hollow carbon microspheres from hollow phenolic microspheres have good potential to be attractive functional fillers for carbon foams [38]. It shows a significant improvement in the fracture toughness of syntactic foam at 20–30 vol.% microspheres [39]. Activated carbon reinforced microcellular phenolic foams are the incipient in heat shielding application as they can be effectively used for thermal shielding due to the low thermal conductivity, density, and reliable compressive strength [40].

(a)

(b)

Figure 3: (a) 15-mm thick silica fibre-reinforced phenolic syntactic foam. (b) Photograph of phenolic foam after aerothermal test [22].

Elastomeric Matrix

Elastomeric based heat shielding materials is a novel concept in the ablative composites due to their excellent ablation resistance, char yield, and high strain rate [41]. The high strain rate performance with excellent dimensional flexibility significantly reduces the induced thermal stresses during thermal expansion of composites. The elastomers show high resistance against rapid removal of char by mechanical shear and spallation during aeroheating loads [42]. Silicone and Nitrile rubber based ablative materials are the initial developments in the elastomeric matrix system due to their high strain rate with reduced thermal stresses. Silicone based elastomeric ablators show better thermal insulation characteristics due to the formation of siliceous char, which is an inert char layer having excellent thermal stability. Among the other polymeric insulators, elastomeric ablators show high density, limited shelf life, and inferior thermal resistance, which limit their wide applications as thermal insulators. Ethylene propylene diene monomer (EPDM) rubber based matrix system is a novel approach in the elastomeric heat shielding materials due to its high oxidation resistance with excellent low temperature properties [43]. EPDM matrix system can be effectively strengthened through reinforcement of glass, carbon, Kevlar, and polysulfonamide fibers. The low density and high thermal capacity with excellent fire resistant of these synthetic fibers effectively improve the thermochemical properties of elastomeric ablative composites [44]. Flame retardant additives such as ammonium polyphosphate are used as an efficient method to enhance thermal insulation properties of the shielding composites.

Recently, Jia et al. have studied the reinforcement effects of polysulfonamide fibers in the EPDM matrix as a thermal insulation fibers and also noticed the enhancement in thermal stability due to the presence of additional sulfone group ($-SO_2-$) in the main chain of the sulfonamide fibers [45]. Similarly, Natali et al. have studied the reinforcement effects of Kynol fibers, made by acid-catalyzed cross-linking of melt-spun Novolac and noticed that Kynol fiber shows excellent thermal properties as compared to Kevlar fibers. Figures 4(a) and 4(b) represent the postburning images of the EPDM based Kynol tested materials and Figure 4(c) illustrates the SEM image of the burnt surface of EPDM/Kynol samples [23]. Since the thermal insulation is inversely proportional to the density of the materials, low density elastomeric ablators need to be designed for a better ablation performance. Reinforcement of additives with a density gradient is an effective approach for the development of a light weight ablator without compromising on the ablation performance. In the recent period, hybrid ablative composites are a topical concept in which two or more fibers reinforced in the matrix system for optimized balance properties in the thermal insulation system. Regarding this, Ahmed and Hoa have studied the effect of a chopped carbon fiber (CCF) and aramid fiber on pulp as hybrid reinforcement for ethylene propylene diene monomer (EPDM) along with ammonium polyphosphate (AP) flame retardant agent and noticed improvement in the ablation performance with an improved thermomechanical performance [46].

(a)

(b)

(c)

Figure 4: ((a) and (b)) Postburning images of the EPDM based Kynol tested materials. (c) SEM image of the burnt surface of EPDM/Kynol samples [23].

NANO FILLER ENABLED HYBRID COMPOSITE

Nanoparticles (or nanopowder or nanocluster or nanocrystal) are the microscopic particles which are having exceptional high surface area to volume ratio that makes them an effective filler reinforcement for ablative composites. Reinforcing effects of the nanoparticle mainly depend on the crystalline microstructure and effective interfacial bonding with the parent matrix [47]. Nanofiller embedded ablative composites show enhancement in thermal degradation and ability to resist high-temperature erosive aerodynamic shear forces. This is mainly due to its high thermal stability, low thermal conductivity, and high adhesive bonding with the parent matrix. Adhesive bonding is mainly related to the improvement in interlaminar shear strength (ILSS) in hybrid composite structure [24]. Recent research works cover the development of carbon nanofibers, nanoclay, and CNT in phenolic resin for better ablation resistance. Dash et al. studied the mechanical characterization of a Red Mud filled hybridized composite and noticed improvement in flexural and tensile properties [48]. Generally, the dispersion of the nanoparticle in the fiber reinforced composite system is in the range of 1–4 wt% [49]. Beyond the percolation threshold, mechanical and ablation resistance of the composite decreased due to the agglomeration of the nanofillers. Uniform dispersion of nanofillers effectively improves the structural integrity of the char layer during ablation. Inorganic fillers are found as more suitable materials for thermal insulations due to the presence of thermal barrier properties and surface morphology [50].

Organic Fillers

Higher aspect ratio and inherent thermomechanical properties are the key factors of organic fillers for their effective usage as reinforcing agent to protect the carbon fibers and matrix from aerodynamic forces. Organic fillers include single wall carbon nanotube (SWCNT), multiwall carbon nanotube (MWCNT), carbon nanofiber, carbon black, and graphene. During ablation, MWCNT in the char layer reemits the incident heat radiation in to gas phase. This reemission decreases the heat transfer rate into the inner layers of the composite panels and reduces the endothermic pyrolysis [51]. So the presence of MWCNT leads to an improvement in the thermal insulation and reduction in mass loss. Tirumali et al. studied different morphological nanofibers (stacked-coin-type) and their morphological shape variation with respect to various mechanical and chemical treatments [47]. Vapour-grown carbon nanofibers have the potential to enhance the structural properties of the char due to its high aspect ratios and fine diameter ranging from 15 nm to 100 nm [46]. In a review by Tirumali et al. on Epoxy composites of Graphene Oxide (GO), the morphological characteristics of Graphene Oxide and Graphene were studied and illustrated that platey like morphology of GO/Graphene enhanced their reinforcement effect on the polymeric system [47]. Figure 5 shows the charred surface of the MWCNT dispersed in carbon-carbon composite. 2 wt% of filler addition makes a thermal barrier protection layer over the carbon fiber and avoids its peel off from the bottom layer.

(a)

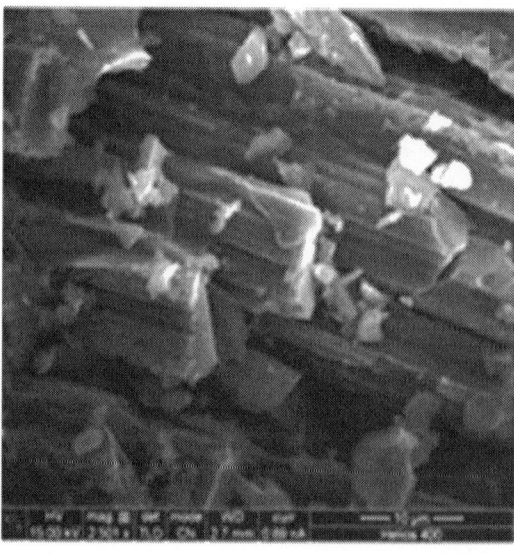

(b)

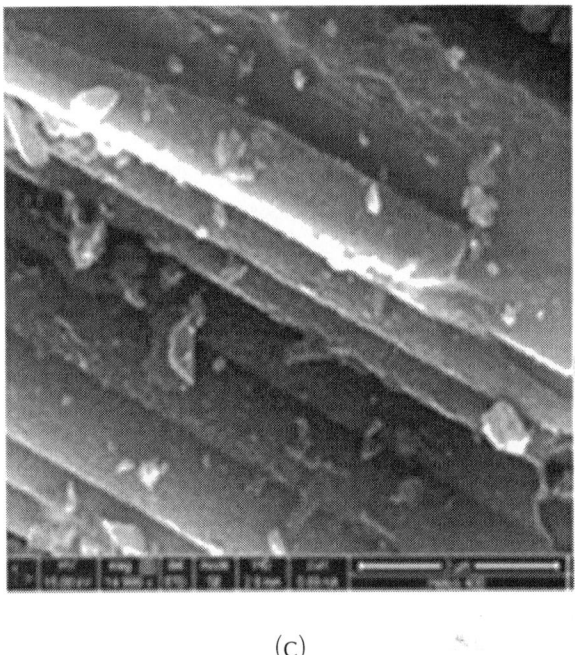

(c)

Figure 5: SEM images of (a) 0.5 wt%, (b) 1 wt%, and (c) 2 wt% MWCNTs ablative nanocomposite samples [18].

Inorganic Fillers

The inorganic nanofillers are the prominent filler reinforcement in the composite material for heat shielding applications. These nanofillers, such as montmorillonite (MMT) nanoclay, fly ash, nanosilica, zirconia diboride, calcium carbonate, and barium sulphate, are being used for their reinforcing ability to enhance the tribological and thermal properties of the polymer matrix. Microsized fillers require large amount of filler loading for a uniform char layer over the surface [47]. Nonuniform dispersion and large void region in microsize fillers create a local nonuniform erosion rate with a rough surface and hence are prone to erosion. Nanofillers show a significant reinforcing effect compared to microsized fillers due to their higher specific surface area and volume effects. The in-organic fillers melt during high temperature exposure and form a viscous melt layer over the burned surface. This layer acts as a protective antioxidation barrier and radiative cooling system in the high temperature burned surface. Molten fused silica reacts with char and produces a SiC phase. Silica melt layer absorbs significant amount of heat by an endothermic process during its phase transformation and enhances the cooling rate [52]. Srikanth et al. studied the effect of nanosilica in carbon phenolic composites and identified an improvement in the ablation resistance along with reduced thermal conductivity and improved interlaminar shear strength [52]. Xiao et al. have reported that addition of zirconium diboride to carbon phenolic composite significantly improves the ablation resistance during the fire exposure [53]. Polyhedral oligomeric silsesquioxane is reported to be efficient filler for reducing thermal conductivity and ablation resistance in the carbon phenolic composites [53]. Figure 6(a) shows the eroded carbon fibers during plasma arc treatment, while Figure 6(b) shows the protective layer of the melt nanosilica over the carbon fibers.

(a)

(b)

Figure 6: (a) Ablated surface of blank C–Ph composite, (b) 2.0 wt.% nano-silica C–Ph after ablation [24].

Inorganic Nanocoating on Carbon Fibers

Surface coating on the reinforcement fiber is the new development in fiber reinforced composite system. The aim of coating on the reinforced fibers is to improve the oxidation resistance and adhesive bonding of the carbon fiber with surrounding matrix. Optimum coating thickness is significant constraint in order to obtain a good compressibility of the hybrid composite system, which maintains the fiber reinforcement volume (%) in the composite system. Srikanth et al. studied the effect of zirconium oxide coating (thickness 700–900 nm) on the carbon fabric and found the reduction in thermal conductivity and the rear face temperature. Surface coating causes the reduction in the effective interfacial area between the fiber and the matrix which leads to a decrement in flexural modulus and interlaminar shear strength [54]. Thermally sprayed tungsten carbide and silicon coating are an effective method for effective thermal barrier shielding [55]. Recently, Ryu has fabricated C/SiC functionally graded coating for reducing the thermal residual stresses between carbon composites and SiC coating layer. As per his observation SiC rich compositional layer in the Functionally Graded material (FGM) shows effective release of the thermal stress and increment in oxidation resistance [56].

FUNCTIONALLY GRADED COMPOSITES

Functionally graded composite material (FGM) is composed of two different phases in which volume fraction changes gradually along at least one dimension of the solid. Functionally graded nanocomposite features layer by layer variation in the filler composition and their concentration density profiles in order to realize a designed functionality. Effective design of the various layers in FGM makes minimum heat diffusion through thickness with maximum thermal insulation [57]. Functionally graded materials are considered to have smooth spatial variation of microstructure and homogenized material properties [26]. This structural variation improves the compressive strength and there is enhancement in the fracture and fatigue strength. Spatial variation of microstructure significantly minimizes the thermal delamination which takes place through the difference in the thermal expansion coefficient of the two materials [58]. The distribution of residual thermal stresses is very important in the soundness of FGM composite. Figure 7 shows the SEM image of the linearly continuous gradient microstructure of carbon-based SiC/C FGM. It signifies that smooth gradient interface in the FGM ensures narrow transition from one material to another and decreases the thermal residual stresses. Srikanth et al. developed a composite panel in which the top layer consists of carbon nanotube (CNT) hybrid carbon phenolic composite and in the bottom layer Zirconium Oxide hybrid C–Ph composite. Presence of the CNT in the composite enhances the flexural and shear strength of the composite, but during charring composite loses its structural integrity due to the improvement in thermal conductivity of the CNT [50]. Powder metallurgical approach including plasma spraying method has been widely used in the fabrication of thermal stress relief type of functionally graded materials [59]. Liu et al. reported the modelling study of thermal stress in SiC/C functionally graded composite by finite element numerical models and noticed the decrement in thermal stresses with increasing the layer number. The optimum intermediate graded layer and pure SiC layer thickness was 3-4 mm and 0.5–1 mm [60]. Bafekrpour et al. studied the flexural properties of the carbon nanofiber/phenolic functionally graded composite by varying the compositional gradient of carbon nanofiber and noticed the improvement in flexural properties of FGM [61].

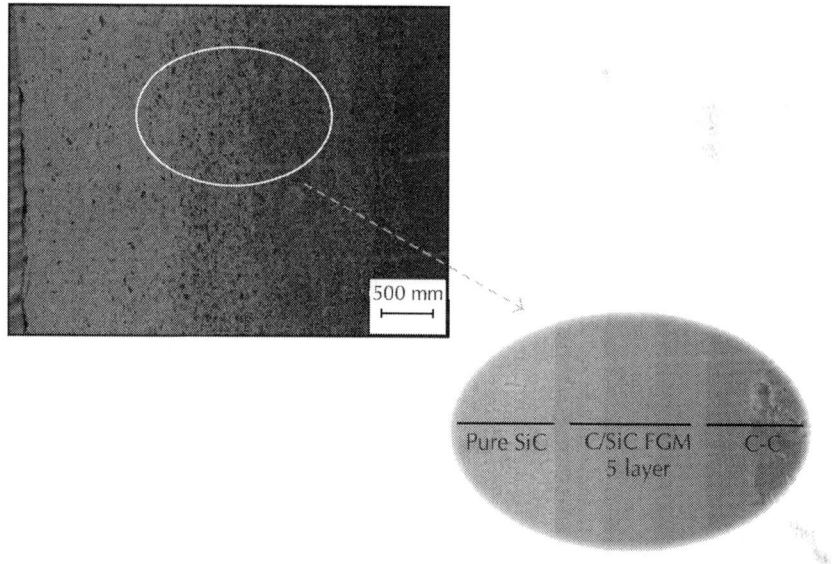

Figure 7: SEM image of the linear continuous gradient microstructure of carbon-based SiC/C FGM and their magnified intermediate transition layers [25].

ABLATIVE MECHANISMS

The ablation mechanism of the carbon-carbon composite in nozzle inserts is a heterogeneous chemical interaction between the propellant combustion products and nozzle materials [62]. It is a thermal protection mechanism in which high temperature radiant energy from the combustion chamber acts on the material surface and dissipated through a series of endothermic reaction processes (thermochemical, thermophysical, and thermomechanical). When the throat surface temperature is less than of 2000–2500 K, ablation mechanism is dominated by chemical kinetics. The temperature at 2000–2500 K, the reaction is significantly controlled by diffusion rate [62]. The graphite or carbon-carbon materials validate a superior advantage since their sublimation at 1 atm was about 4000 K and it increased with the increase in pressure. Ablation concedes through a combined action of heat transfer, chemical reactions, and fluid flow and finally the material is consumed [63]. Figure 8 shows the ablation mechanism in the carbon based polymer matrix ablatives and illustrates the combined thermomechanical decomposition phenomena during ablation process. During the space craft propulsion, hypersonic flow of hot air creates an external boundary layer of low pressure over the ablation surface. The circulation of the dissipated heat occurs within this boundary and passes the heat away from the material surface. This phenomenon led to an added advantage for an efficient thermal shielding mechanism [50]. Normally, the thermochemical degradation of outer layer of the carbon-carbon composite takes place through convectional heat transfer in which pyrolyzing layer diffuses towards the heated area of the shield. In the nozzle design, the pressure and temperature of the combustion gases at the throat could not match the condition for sublimation, so sublimation was neglected as an ablation mechanism for the throat erosion [51]. Interface region of the fiber/matrix and the defective regions of the composite are more prone to ablation. This is due to the lower activation energy and high reactivity of the defective region in the composite system [64, 65]. The ablation always developed along the direction of interface-to-fiber and interface-to-matrix due to the high oxidization tendency of the interface region. Donghwan has explained ablation mechanism by an erosive phenomenon, who reported that decomposition of the ablative materials takes place through thermal oxidation during fire exposure at high temperature of 2000–3000°C, resulting in the formation of char and peels off from parent layer by the combustion flame [12]. These typical ablation characteristics suggest that super lightweight carbon ablators require high thermal degradation temperature and less dense microstructure to perform as an efficient heat shielding material [66]. Figure 9 shows the schematic of parameters influencing ablation resistance of the carbon/carbon composite.

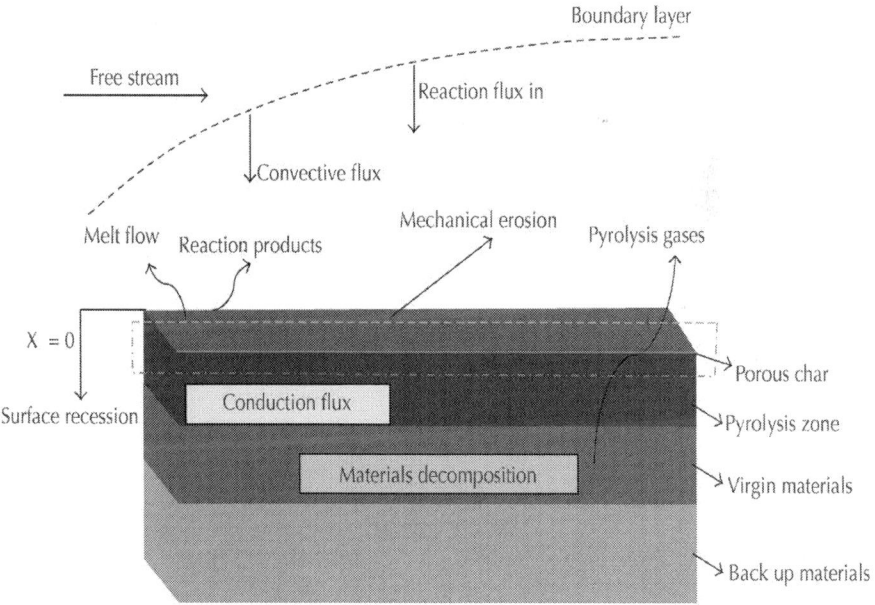

Figure 8: Pyrolysis mechanism carbon based polymer matrix [26].

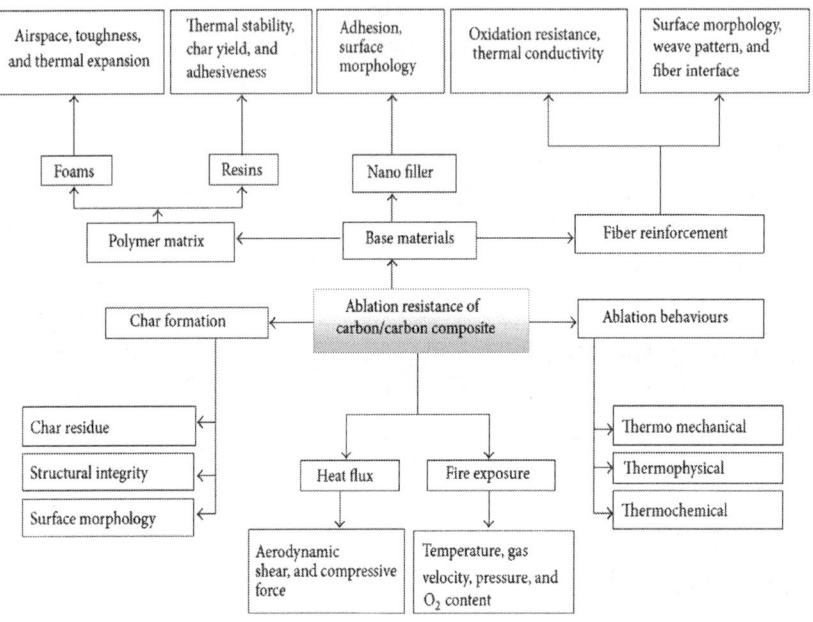

Figure 9: Parameters affecting ablation resistance of phenolic matrix ablatives.

Thermal Decomposition

In case of carbon-carbon polymer composites, decomposition takes place through the viscous softening, melting, decomposition, and volatilization [67]. So a meticulous study over the decomposition characteristics to observe the real material degradation phenomenon is essential. The endothermic pyrolysis mechanism during the ablation process depends on the temperature, heat flux, time duration of the fire, and type of aerodynamic stress (e.g., tension, compression, bending, and torsion) [68]. In the initial stage of the heat flux, heat transfer takes place through pure conduction and hence it causes rise in temperature. At a particular temperature, the pyrolysis reaction (thermal degradation) takes place followed by the elimination of pendant aromatic rings and retention of all aromatic carbons [69]. This reaction causes the formation of a thermally cross-linked intermediate structure of new liquid or gas phases. In the decomposition process, resin evaporation takes place and causes development of pressure in the composite due to the penetration of evaporated gases through the fiber laminates. The excess pressure creates volume expansion in the remaining matrix and intense material degradation rate causes carbon laminates to lose their structural integrity. During thermal degradation thermal conductivity (K), specific heat (C), and density () of the composite change [70]. Fiber decomposition behaviour depends on the fiber reinforcement density area. The fiber degrades to cone shape fiber during ablation in high fiber density area. In the low fiber density area, the fibers get weakened during ablation and peel off from the matrix due to aerodynamic stresses [71].

Pyrolysis Mechanism of Carbon Based Polymer Matrix

Pyrolysis of carbon based polymer matrix takes place through a three stage process. First stage of the pyrolysis mechanism demonstrates the formation of additional cross-links as a result of two condensation reactions between functional groups of the cured phenolic. At second stage, breakage of the cross-links occurred followed by the evolution of methane, hydrogen and carbon monoxide. Similarly, final stage causes stripping of the hydrogen atoms from the ring structure and evolution of hydrogen gas. In pyrolysis reactions, the phenolic matrix converts into amorphous carbon with less structural integrity. This is due to the complete elimination of the noncarbon species and char of coalesced carbon rings formation during pyrolysis. Evolution of gas during the pyrolysis mechanism is controlled by controlling the heating rate of the composite [72]. Clarity in the physiochemical reaction in pyrolysis mechanism leads to optimization of the factor of safety during the design of the composite heat shields. Optimized factor of safety leads to an efficient design of heat shields with least material wastages and max thermal protection.

calculating the temperature distribution in composites. This one-dimensional nonlinear equation accounts for the energy transfer through heat conduction, pyrolysis mechanism, and diffusion of decomposition gases which is shown in the following [82]:

$$\rho C_p \frac{\partial T}{\partial t} = k_\perp \frac{\partial^2 T}{\partial x^2} + \frac{\partial k_\perp}{\partial x}\frac{\partial T}{\partial x} - \dot{m}_g C_{p(g)}\frac{\partial T}{\partial x}$$

$$- \frac{\partial \rho}{\partial t}\left(Q_p + h_c - h_g\right),$$

(1)

where h_c and h_g represent the enthalpy of solid and gas phase.

$$h_c = \int_{T_\infty}^{T} C_p dT,$$

$$h_g = \int_{T_\infty}^{T} C_{p(g)} dT.$$

(2)

The first term one the right hand side of (1) accounts for the heat conduction through thickness of the composite laminate. The second term accounts for the changes in the thermal conductivity with increasing temperature. The third term accounts for the internal convection of thermal energy due to the flow of volatile gases from pyrolysis reaction. Similarly the last term accounts for the temperature change in the composite due to heat generation or consumption resulting from endothermic reaction of the matrix material. Equation (1) presumes a one-dimensional material system with the heat transfer occurring only through thickness direction. Equation (1) is expanded to analyse two- and three-dimensional systems with large amount of empirical data on the thermal and decomposition properties of the composite. Modification of the Henderson's thermal model has been done by many research groups. In the recent period, Gibson et al. slightly modified the Henderson model to include the decomposition reaction rate of the polymer matrix. This model can be used to calculate the mass loss and extent of charring during decomposition [83]. Gibson et al. recently proposed a simplified model by avoiding the use of the Arrhenius decomposition model. The new model "apparent thermal diffusivity (ATD)," involves expression of the thermal diffusivity of the composite as a function of temperature. ATD takes into account the decomposition of the resin, which is endothermic, as well as consequently changes specific heat capacity and thermal conductivity of the composite. The rate dependence of the decomposition behaviour is neglected, with the decomposition state at any temperature being calculated directly from the TGA curve for the polymer matrix.

Modelling Thermal Properties of Decomposing Composites

Thermal decomposition of the composite material depends on thermal and physical properties such as density, thermal conductivity, specific heat capacity, and gas permeability of both virgin composite and its char residue [82]. Temperature dependence of these properties is important for the prediction of thermal response of the composite during thermal analysis. During thermal decomposition and pyrolysis, density of composite varies significantly. Arrhenius decomposition kinetics is the common approach for finding the change in density with time. The density change in a single stage decomposition reaction is given by the following:

$$\frac{\partial \rho}{\partial t} = -(\rho_v - \rho_c)\left[\frac{\rho - \rho_c}{\rho_v - \rho_c}\right]^n Ae^{(-E/RT)}$$

(3)

Thermal properties of the fully decomposed phase (char) are difficult to obtain due to unavailability of the measured data. So during thermal analysis the properties of the decomposing composite are assumed to be dependent on the relative mass fractions of the virgin composite and its char. The mass fraction of virgin material in a decomposing composite can be calculated as

$$F = \left[\frac{\rho - \rho_c}{\rho_v - \rho_c} \right]$$

(4)

By considering the density changes as a variable element in the matrix decomposition, thermal conductivity of the decomposing composite can be defined as follows:

$$k(T) = F \cdot k_v(T) + (1 - F) \cdot k_c(T)$$

(5)

Thermal conductivity depends on the thermal response of composites in fire, considered to be an important parameter for thermal analysis of the composite structure. Henderson and Wiecek have reported that the specific heat capacities for the virgin composite and char depend on the temperature. The specific heat capacity is determined using the following:

$$C_p(T) = F \cdot C_{p(v)}(T) + (1 - F) \cdot C_{p(c)}(T)$$

(6)

By using different parameters, thermal decomposition behaviour of the ablative composite can be well analyzed. However, fire-induced damage such as fibre-matrix debonding, intraply matrix cracking, and fibre damage degrade the thermal properties of the composites. Delaminations of the laminates affect the reduction in thermal conductivity of the virgin composite due to the formation of air gap between debonded ply layers [77]. So there is a requirement for unified damage model by incorporating the influence of various damages in the composite structures during thermal decomposition.

Mathematical Model of Phenolic Impregnated Carbon Composite

Pyrolysis mechanism of the carbon-carbon composite is as same as other ablative materials in terms of thermal decomposition, char morphology, and delamination. In case of phenolic resin, the matrix decomposition takes place at a particular temperature followed by the char layer ablation at a little higher temperature

[78]. Thermal response of carbon-carbon composite at various zones (virgin material zone, pyrolysis zone, and porous char layer zone) is different and the thermal analysis needs to consider for this variation in thermal response.

The main assumptions during the ablation models are as follows.

- Energy transfer through mass diffusion is neglected.
- Mass transfer takes place mainly through the movement of pyrolysis gases.
- Pyrolysis gases considered "ideal gases" and their properties are constant.
- The specific heat capacity of the composite is a mass weighted average of the relative mass fractions of polymer, char, and fiber remaining in the composite.
- Heat conductivity coefficient mainly depends on the temperature variation.
- Thermal decomposition is a single stage process and considered as first-order reaction.

Heating region in the ablative composite is composed of the ionized air that has dissociated. Mass and energy transfer of the volatile products result in complex chemical reactions at the surface and within the boundary layer. The balance between the convective energy, chemical energy, and net radiation at the ablative composite surface are as follows [79]:

$$\rho_e\,\mu_e C_H\left(H_r - h_{ew}\right) + \rho_e\mu_e C_M$$

$$\times \left[\sum\left(Z_{ie}^n - Z_{iv}^n\right)h_i^{Tv} - B'h_w\right] + \dot{m}_c h_c + \dot{m}_g h_g$$

$$+ \alpha_w q_{\mathrm{rad}} - F\sigma\varepsilon_w T_w^4 - q_{\mathrm{cond}} = 0. \tag{7}$$

While considering the reaction inside the ablative composite, the effect of surface recession and pyrolysis gas is to be taken into account because, rate of recession of the material thickness affects the conduction and thermal capacitance and escaping of the volatile gas, significantly affecting the thermal decomposition [79].

It can be represented as follows:

$$\rho c_p \frac{\partial T}{\partial t} = \frac{1}{A} \frac{\partial}{\partial x} \left\{ kA \frac{\partial T}{\partial x} \right\} + \left(h_g - \bar{h} \right) \frac{\partial p}{\partial t}$$

$$+ \dot{s} \rho c_p \frac{\partial T}{\partial x} + \frac{\dot{m}}{A} \frac{\partial h_g}{\partial x}.$$

(8)

   In thermal shielding composites, there is a constant requirement of mechanistic based models for analyzing the temperature distribution, damage, softening, and failure of the ablative composite. However, surplus empirical data in the mechanistic based model reduce the reliance of the thermal analysis. Extensive studies need to be performed on the ablation mechanism in order to enhance the accuracy and prediction in thermal analysis.

CONCLUSIONS

Ablative materials are efficient thermal insulators due to their uniqueness in accommodating high heat flux and resistance against the heat diffusion. Enhancement in the ablation properties is achievable through matrix modification, improving the surface texture of the fiber reinforcement and dispersion of various nanofillers. Thermal decomposition resistance, ablation resistance, and char retention of the phenolic enhanced with modification of high thermal resistive elements such as Boron, Titanium, and Molybdenum. Elastomeric matrix based heat shielding composites show excellent thermal insulation properties due to their excellent ablation resistance, char yield, and high strain rate. Reinforcing elastomeric matrix system with the synthetic fibers leads to a better ablation properties and enhancement in thermal insulation behaviour. Organic fillers, such as carbon nanotubes, carbon nanofiber, graphite oxide, and graphene, have high potentials in both thermal insulation and char structural integrity. Thermal resistive behaviour of the nanofillers is based on their morphological features and inherent physiochemical characteristics. Inorganic fillers like nanosilica, montmorillonite (MMT) nanoclay, and glass balloons, have higher thermal barrier property. Nanofillers having hollow sphere morphology (Fly ash, Glass balloons) have the effective resistance against heat diffusion rate due to the air space volume in the hollow sphere. Oxidation and aerodynamic erosion resistance of the carbon fiber is achieved by improving the adhesive bonding of the fiber and inorganic coating over the carbon fiber. The main challenge lies in the thermal analysis of the composite materials during ablation. Assessment of the composite properties during the thermal degradation is quite challenging. Combined thermal-structural analysis methodology is needed for carbon-carbon composite materials to deal with the varying working environment and expected service requirement. Furthermore, mechanical damage during the fire exposure, such as delamination cracking and skin-core debonding, needs more attention in the future thermal design of ablation materials.

ACKNOWLEDGMENTS

The authors would like to thank Dr. Prahlada, Vice Chancellor of Defence Institute of Advanced Technology (DIAT), for his support and encouragement. In addition, we are grateful to Group. Captain. T Manoj, Dr. Renuka Gonte, B. N. Sahoo, Premika. G, and Colonel Vijay Kumar DIAT (DU) for proofreading the paper and providing many useful comments.

REFERENCES

1.   T.-C. Chen and C.-C. Liu, "Inverse estimation of heat flux and temperature on nozzle throat-insert inner contour," International Journal of Heat and Mass Transfer, vol. 51, no. 13-14, pp. 3571–3581, 2008.

2.   Y. Tong, S. Bai, H. Zhang, and Y. Ye, "Laser ablation behavior and mechanism of C/SiC composite," Ceramics International, vol. 39, no. 6, pp. 6813–6820, 2013.

3.   A. P. Mouritz, S. Feih, E. Kandare et al., "Review of fire structural modelling of polymer composites," Composites A, vol. 40, no. 12, pp. 1800–1814, 2009.

4.   Y. Chen, P. Chen, C. Hong, B. Zhang, and D. Hu, "Improved ablation resistance of carbon-phenolic composites by introducing zirconium diboride particles,"Composites B, vol. 47, pp. 320– 325, 2013.

5.   T. K. Kuo and S. Keswani, "A comprehensive theoretical model for carbon-carbon composite nozzle recession," Combustion Science and Technology, vol. 42, no. 3-4, pp. 145–164, 1985.

6.   P. Thakre and V. Yang, "Chemical erosion of carbon-carbon/ graphite nozzles in solid-propellant rocket motors," Journal of Propulsion and Power, vol. 24, no. 4, pp. 822–833, 2008.

7.   T. Windhorst and G. Blount, "Carbon-carbon composites: a summary of recent developments and applications," Materials and Design, vol. 18, no. 1, pp. 11–15, 1997.

8.   M. H. Al-Saleh and U. Sundararaj, "Review of the mechanical properties of carbon nanofiber/polymer composites," Composites A, vol. 42, no. 12, pp. 2126–2142, 2011.

9.   K. A. Trick and T. E. Saliba, "Mechanisms of the pyrolysis of phenolic resin in a carbon/phenolic composite," Carbon, vol. 33, no. 11, pp. 1509–1515, 1995.

10. C. Hong, J. Han, and X. Zhag, "Novel phenolic impregnated 3- D fine-woven pierced carbon fabric composites: microstructure and ablation behavior," Composites B, vol. 43, no. 5, pp. 2389– 2394, 2012.

11. S. Chand, "Review carbon fibers for composites," Journal of Materials Science, vol. 35, no. 6, pp. 1303–1313, 2000.

12. C. Donghwan, "A microstructural study of the improved ablation resistance of carbon/phenolic composites fabricated using H3PO4-coated carbon fibres," Journal of Materials Science Letters, vol. 15, no. 20, pp. 1786–1788, 1996.

13. R. D. Patton, C. U. Pittman Jr., L. Wang, J. R. Hill, and A. Day, "Ablation, mechanical and thermal conductivity properties

of vapor grown carbon fiber/phenolic matrix composites," Composites A, vol. 33, no. 2, pp. 243–251, 2002.

14. F. W. J. van Hattum, A study of the mechanical properties of vapour grown carbon fibres and carbon fibre-thermoplastic composites, Ph.D. dissertation. Universidade do Minho, 1999, http://repositorium.sdum.uminho.pt/bitstream/1822/282/1/van%20Hattum-Tese%20Doutoramento.pdf.

15. G. G. Tibbetts, M. L. Lake, K. L. Strong, and B. P. Rice, "A review of the fabrication and properties of vapor-grown carbon nanofiber/polymer composites," Composites Science and Technology, vol. 67, no. 7-8, pp. 1709–1718, 2007.

16. K. Lozano, "Vapor-grown carbon-fiber composites: processing and electrostatic dissipative applications," Journal of Management, vol. 52, no. 11, pp. 34–36, 2000.

17. S. R. Dhakate, R. B. Mathur, and T. L. Dhami, "Development of vapor grown carbon fibers (VGCF) reinforced carbon/carbon composites," Journal of Materials Science, vol. 41, no. 13, pp. 4123–4131, 2006.

18. H. Jaeger and T. Behrsing, "The dual nature of vapour-grown carbon fibres," Composites Science and Technology, vol. 51, no. 2, pp. 231–242, 1994.

19. J. S. Tate, S. Gaikwad, N. Theodoropoulou, E. Trevino, and J. H. Koo, "Carbon/phenolic nanocomposites as advanced thermal protection material in aerospace applications," Journal of Composites, vol. 2013, Article ID 403656, 9 pages, 2013.

20. S. G. Advani and C. L. Tucker, "Part B: constitutive equations and flow processing—processing short-fiber systems," in Flow and Rheology in Polymer Composites Manufacturing, S. G. Advani, Ed., vol. 10 of Composite Materials, p. 147, Elsevier Science, 1st edition, 1994.

21. J. Zhao, "Effect of post production processing on dispersion of carbon nanofibers in water," Industrial & Engineering Chemistry Research, vol. 50, no. 3, pp. 1599–1604, 2011.

22. B. John, B. Deependran, G. Joseph, R. C. P. Nair, and K. N. Ninan, "Medium-density ablative composites: processing, characterisation and thermal response under moderate atmospheric re-entry heating conditions," Journal of Materials Science, vol. 46, no. 15, pp. 5017–5028, 2011.

23. M. Natali, M. Rallini, D. Puglia, J. Kenny, and L. Torre, "EPDM based heat shielding materials for solid rocket motors: a comparative study of different fibrous reinforcements," Polymer Degradation and Stability, vol. 98, no. 11, pp. 2131–2139, 2013.

24. D. M. Allison, A. J. Marchand, and R. M. Morchat, "Fire performance of composite materials in ships and offshore structures," Marine Structures, vol. 4, no. 2, pp. 129–140, 1991.

25. S. R. Dhakate, R. B. Mathur, and T. L. Dhami, "Development of vapor grown carbon fibers (VGCF) reinforced carbon/carbon composites," Journal of Materials Science, vol. 41, no. 13, pp. 4123–4131, 2006.

26. R. Lipton, "Design of functionally graded composite structures in the presence of stress constraints," International Journal of Solids and Structures, vol. 39, no. 9, pp. 2575–2586, 2002.

27. A. P. Mouritz and A. G. Gibson, "Fire properties of polymer composite materials," in Solid Mechanics and Its Applications, G. M. L. Gladwell Eds, Ed., pp. 143–163, Springer, Amsterdam, The Netherlands, 1st edition, 2006.

28. C. P. R. Nair, "Advances in addition-cure phenolic resins," Progress in Polymer Science, vol. 29, no. 5, pp. 401–498, 2004.

29. Y. Zhang, S. Shen, and Y. Liu, "The effect of titanium incorporation on the thermal stability of phenol-formaldehyde resin and its carbonization microstructure," Polymer Degradation and Stability, vol. 98, no. 2, pp. 514–518, 2013.

30. C. Luo, W. Xie, and P. E. DesJardin, "Fluid-structure simulations of composite material response for fire environments," Fire Technology, vol. 47, no. 4, pp. 887–912, 2011.

31.   G. Yi and F. Yan, "Mechanical and tribological properties of phenolic resin-based friction composites filled with several inorganic fillers," Wear, vol. 262, no. 1-2, pp. 121–129, 2007.

32.   L. K. Kucner and H. L. McManus, "Experimental studies of composite laminates damaged by fire," in Proceedings of the 26th International SAMPE Technical Conference, vol. 44, pp. 341–353, Paris, France, October 1994.

33.   J. Wang, H. Jiang, and N. Jiang, "Study on the pyrolysis of phenol-formaldehyde (PF) resin and modified PF resin," Thermochimica Acta, vol. 496, no. 1-2, pp. 136–142, 2009.

34.   H. Fan, X. Li, Y. Liu, and R. Yang, "Thermal curing and degradation mechanism of polyhedral oligomeric octa(propargylaminophenyl)silsesquioxane," Polymer Degradation and Stability, vol. 98, no. 1, pp. 281–287, 2013.

35.   D. Wei, D. Li, L. Zhang, Z. Zhao, and Y. Ao, "Study on phenolic resin foam modified by montmorillonite and carbon fibers," Procedia Engineering, vol. 27, pp. 374–383, 2012.

36.   J. Zhou, Z. Yao, Y. Chen, D. Wei, and Y. Wu, "Thermomechanical analyses of phenolic foam reinforced with glass fiber mat," Materials & Design, vol. 51, pp. 131–135, 2013.

37.   H. Shen, A. J. Lavoie, and S. R. Nutt, "Enhanced peel resistance of fiber reinforced phenolic foams," Composites A, vol. 34, no. 10, pp. 941–948, 2003.

38.   L. Zhang and J. Ma, "Effect of coupling agent on mechanical properties of hollow carbon microsphere/phenolic resin syntactic foam," Composites Science and Technology, vol. 70, no. 8, pp. 1265–1271, 2010.

39.   S. Lei, Q. Guo, J. Shi, and L. Liu, "Preparation of phenolicbased carbon foam with controllable pore structure and high compressive strength," Carbon, vol. 48, no. 9, pp. 2644–2646, 2010.

40.   J. Zhou, Z. Yao, Y. Chen, D. Wei, and Y. Wu, "Thermomechanical analyses of phenolic foam reinforced with glass fiber mat," Materials & Design, vol. 51, pp. 131–135, 2013.

41. Z. Jia, G. Li, Y. Yu, G. Sui, H. Liu, and Y. Li, "Effects of pretreated polysulfonamide pulp on the ablation behavior of EPDM composites," Materials Chemistry and Physics, vol. 112, no. 3, pp. 823–830, 2008.

42. C. M. Bhuvaneswari, M. S. Sureshkumar, S. D. Kakade, and M. Gupta, "Ethylene-propylene diene rubber as a futuristic elastomer for insulation of solid rocket motors," Defence Science Journal, vol. 56, no. 3, pp. 309–320, 2006.

43. W. K. Ho, J. H. Koo, and O. A. Ezekoye, "Thermoplastic polyurethane elastomer nanocomposites: morphology, thermophysical, and flammability properties," Journal of Nanomaterials, vol. 2010, Article ID 583234, 11 pages, 2010.

44. A. S. Deuri and A. K. Bhowmick, "Ageing of rocket insulator compound based on EPDM," Polymer Degradation and Stability, vol. 16, no. 3, pp. 221–239, 1986.

45. X. Jia, G. Li, Y. Yu et al., "Ablation and thermal properties of ethylene-propylene-diene elastomer composites reinforced with polysulfonamide short fibers," Journal of Applied Polymer Science, vol. 113, no. 1, pp. 283–289, 2009.

46. A. F. Ahmed and S. V. Hoa, "Thermal insulation by heat resistant polymers for solid rocket motor insulation," Journal of Composite Materials, vol. 46, no. 13, pp. 1549–1559, 2012.

47. M. Tirumali, K. Balasubramanian, and A. Kumaraswamy, "Epoxy composites of graphene oxide (GO): a review," in Proceedings of the IEEE International Conference on Research and Development Prospects on Engineering and Technology (ICRDPET '13), vol. 1, p. 94, Nagapattinam, India, March 2013.

48. A. K. Dash, D. N. Thatoi, and M. K. Sarangi, "Analysis of the mechanical characteristics of a red mud filled hybridized composite," in Proceedings of the International Conference on Frontiers in Automobile and Mechanical Engineering (FAME '10), pp. 8–11, November 2010.

49. M. M. Zurale and S. J. Bhide, "Properties of fillers and reinforcing fibers," Mechanics of Composite Materials, vol. 34, no. 5, pp. 463–472, 1998.

50. C. Luo and P. E. DesJardin, "Thermo-mechanical damage modeling of a glass-phenolic composite material," Composites Science and Technology, vol. 67, no. 7-8, pp. 1475–1488, 2007.

51. H. L. McManus, "Prediction of fire damage to composite aircraft structures," in Proceedings of the 9th International Conference on Composite Materials (ICCM-9 '93), vol. 58, pp. 929–936, Madrid, Spain, 1993.

52. I. Srikanth, A. Daniel, S. Kumar et al., "Nano silica modified carbon-phenolic composites for enhanced ablation resistance," Scripta Materialia, vol. 63, no. 2, pp. 200–203, 2010.

53. J. Xiao, J. Chen, H. Zhou, and Q. Zhang, "Study of several organic resin coatings as anti-ablation coatings for supersonic craft control actuator," Materials Science and Engineering A, vol. 452-453, pp. 23–30, 2007.

54. I. Srikanth, N. Padmavathi, S. Kumar, P. Ghosal, A. Kumar, and C. Subrahmanyam, "Mechanical, thermal and ablative properties of zirconia, CNT modified carbon/phenolic composites," Composites Science and Technology, vol. 80, pp. 1–7, 2013.

55. K.-Z. Li, X.-T. Shen, H.-J. Li, S.-Y. Zhang, T. Feng, and L.- L. Zhang, "Ablation of the carbon/carbon composite nozzlethroats in a small solid rocket motor," Carbon, vol. 49, no. 4, pp. 1208–1215, 2011.

56. Y. Xu, W. Zhang, D. Chamoret, and M. Domaszewski, "Minimizing thermal residual stresses in C/SiC functionally graded material coating of C/C composites by using particle swarm optimization algorithm," Computational Materials Science, vol. 61, pp. 99–105, 2012.

57. V. A. Rozenenkova, N. A. Mironova, S. S. Solntsev, and S. V. Gavrilov, "Ceramic coatings for functionally graded high-

temperature heat-shielding materials," Glass and Ceramics, vol. 70, no. 1-2, pp. 26–28, 2013.

58. C. C. Ma and Y. T. Chen, "Theoretical analysis of heat conduction problems of nonhomogeneous functionally graded materials for a layer sandwiched between two half-planes," Acta Mechanica, vol. 221, no. 3-4, pp. 223–237, 2011.

59. H. Guo, K. A. Khor, Y. C. Boey, and X. Miao, "Laminated and functionally graded hydroxyapatite/yttria stabilized tetragonal zirconia composites fabricated by spark plasma sintering," Biomaterials, vol. 24, no. 4, pp. 667–675, 2003.

60. G. Zhang, Q. Guo, K. Wang et al., "Finite element design of SiC/C functionally graded materials for ablation resistance application," Materials Science and Engineering A, vol. 488, no. 1-2, pp. 45–49, 2008.

61. E. Bafekrpour, C. Yang, M. Natali, and B. Fox, "Functionally graded carbon nanofiber/phenolic nanocomposites and their mechanical properties," Composites A, vol. 54, pp. 124–134, 2013.

62. J. H. Koo, H. Stretz, A. Bray et al., "Nanostructured materials for rocket propulsion system: recent progress," in Proceedings of the 44th AIAA/ASME/ASCE/AHS/ASC Structures, Structural Dynamics, and Materials Conference, p. 1769, Virginia, Va, USA, April 2003.

63. J. S. Tate, S. Gaikwad, N. Theodoropoulou, E. Trevino, and J. H. Koo, "Carbon/phenolic nanocomposites as advanced thermal protection material in aerospace applications," Journal of Composites, vol. 2013, Article ID 403656, 9 pages, 2013.

64. P. Thakre and V. Yang, "Chemical erosion of carbon-carbon/graphite nozzles in solid-propellant rocket motors," Journal of Propulsion and Power, vol. 24, no. 4, pp. 822–833, 2008.

65. A. J. Goupee and S. S. Vel, "Transient multiscale thermoelastic analysis of functionally graded materials," Composite Structures, vol. 92, no. 6, pp. 1372–1390, 2010.

66. J. Kim, S. W. Lee, and S. K. Won, "Time-to-failure of compressively loaded composite structures exposed to fire," Journal of Composite Materials, vol. 41, no. 22, pp. 2715–2735, 2007.

67. C. Luo and P. E. Des Jardin, "Thermo-mechanical damage modeling of a glass-phenolic composite material," Composites Science and Technology, vol. 67, no. 7-8, pp. 1475–1488, 2007.

68. R. Palaninathan, "Behavior of carbon-carbon composite under intense heating," International Journal of Aerospace Engineering, vol. 2010, Article ID 257957, 7 pages, 2010.

69. T. Lippert and J. T. Dickinson, "Chemical and spectroscopic aspects of polymer ablation: special features and novel directions," Chemical Reviews, vol. 103, no. 2, pp. 453–486, 2003.

70. W. Xie and P. E. DesJardin, "An embedded upward flame spread model using 2D direct numerical simulations," Combustion and Flame, vol. 156, no. 2, pp. 522–530, 2009.

71. G. Pulci, J. Tirillo, F. Marra, F. Fossati, C. Bartuli, and T. Valente, ` "Carbon-phenolic ablative materials for re-entry space vehicles: manufacturing and properties," Composites A, vol. 41, no. 10, pp. 1483–1490, 2010.

72. V. Srebrenkoska, G. Bogoeva-Gaceva, and D. Dimeski, "Composite material based on an ablative phenolic resin and carbon fibers," Journal of the Serbian Chemical Society, vol. 74, no. 4, pp. 441–453, 2009.

73. L. Chen, C. Luo, J. Lua, and J. Shi, "A direct coupling approach for fire and composite structure interaction," in Proceedings of the 17th International Conference on Composite Materials (ICCM-17 '09), Edinburgh International Convention Centre (EICC), Edinburgh, UK, July 2009.

74. J. Florio Jr., J. B. Henderson, F. L. Test, and R. Hariharan, "A study of the effects of the assumption of local-thermal equilibrium on the overall thermally-induced response of a decomposing, glass-filled polymer composite," International

Journal of Heat and Mass Transfer, vol. 34, no. 1, pp. 135–147, 1991.

75. L. Torre, J. M. Kenny, and A. M. Maffezzoli, "Degradation behaviour of a composite material for thermal protection systems part I-experimental characterization," Journal of Materials Science, vol. 33, no. 12, pp. 3137–3143, 1998.

76. Y. Hu, X. W. Zhang, and H. You, "Morphology measurement on phenolic-resin/vitreous-silica-fabric ablation composites modified with tetraethoxysilicate and silsesquioxanes," Applied Mechanics and Materials, vol. 333–335, pp. 1934–1937, 2013.

77. A. R. Bahramian, M. Kokabi, M. H. N. Famili, and M. H. Beheshty, "Ablation and thermal degradation behaviour of a composite based on resol type phenolic resin: Process modeling and experimental," Polymer, vol. 47, no. 10, pp. 3661–3673, 2006.

78. A. N. Negovskii, A. V. Drozdov, V. V. Kutanyak et al., "Experimental equipment for the evaluation of the strength characteristics of carbon-carbon composite mateials within the temperature range 20–2200∘ C," Strength of Materials, vol. 31, no. 3, pp. 319–325, 1999.

79. W. Xie and P. E. DesJardin, "A level set embedded interface method for conjugate heat transfer simulations of low speed 2D flows," Computers and Fluids, vol. 37, no. 10, pp. 1262–1275, 2008.

80. J. G. Quintiere, R. N. Walters, and S. Crowley, "Flammability properties of aircraft carbon-fiber structural composite," Technical Report DOT/FAA/AR-07/57, 2007, http://www.fire.tc .faa.gov/pdf/07-57.pdf.

81. G. L. Vignoles, Y. Aspa, and M. Quintard, "Modelling of carboncarbon composite ablation in rocket nozzles," Composites Science and Technology, vol. 70, no. 9, pp. 1303–1311, 2010.

82. J. B. Henderson, J. A. Wiebelt, and M. R. Tant, "Model for the thermal response of polymer composite materials with

experimental verification," Journal of Composite Materials, vol. 19, no. 6, pp. 579–595, 1985.

83. A. G. Gibson, Y.-S. Wu, H. W. Chandler, J. A. D. Wilcox, and P. Bettess, "Model for the thermal performance of thick composite laminates in hydrocarbon fires," Revue de l'Institute Francais du Petrole, vol. 50, no. 1, pp. 69–74, 1995.

Chapter 7

Characterization and Evaluation of Bond Strength of Dental Polymer Systems Modified with Hydroxyapatite Nanoparticles

Vitor César Dumont[1,2], Rafael Menezes Silva[1,2],
Luiz Edmundo Almeida-Júnior[1,2],
Juan Pedro Bretas Roa[3], Adriana Maria Botelho[1],
and Maria Helena Santos[1,2]

[1]Department of Dentistry, Federal University of Vales do Jequitinhonha and Mucuri, Diamantina, Brazil

[2]Advanced Biomaterial Center-BioMat, UFVJM, Diamantina, Brazil

[3]Institute of Science and Technology, UFVJM, Diamantina, Brazil

ABSTRACT

The study modifies an adhesive system with hydroxyapatite nanoparticles as load, characterizes it and evaluates the effectiveness of its bond to dental structure. The middle thirds of healthy premolar tooth crowns were obtained, and each crown was sectioned vertically, resulting in two sections. The sections were divided into ten groups (n = 15), in which resin composite restorations were simulated: (G1E and G1D) conventional adhesive system (SAC); (G2E and G2D) SAC modified with HAP; (G3E and G3D) Primer modified with HAP; (G4E and G4D) monocomponent adhesive system; (G5E and G5D) self-etching adhesive system. The specimens were submitted to the microshear test and characterization technique. There was statistically significant difference (Kruskal-Wallis) between the groups (p < 0.01). G3 presented the highest bond strength to enamel (64.40 MPa, ±7.36) and dentin (39.59 MPa, ±21.46). The majority of specimens were found adhesive fractures. Bond strength to enamel and dentin of the primer modified with HAP of SAC showed higher values.

INTRODUCTION

In Dentistry, one desires a true chemical and micromechanical bond of the restorative material to the tooth structure, even under adverse conditions of humidity and thermal variations in the oral cavity [1]. The development of dental adhesive agents has produced a range of composites classified according to the type of tissue conditioning, and the number of steps for their application [2-4] with direct influence on the bond strength to the dental structure [5].

The bond to dentin is more complex than that to enamel, due to its heterogeneous nature, with a larger organic content and water [6]. Dentin is a tissue that requires a wet bonding technique, since it is formed of hydric components with distinct and variable morphology. This characteristic demands adhesive systems with increasingly hydrophilic formulations, in an attempt to improve the mechanical retention of composites to dentin. There are still few reports on the mechanical performance of amphiphilic adhesive systems which, in addition to a mechanical bond, may promote a strong and lasting chemical bond to dentin [7].

After light activation, the approximation of monomers to establish cross links causes a significant reduction in the volume of the polymer composite. The high coefficient of linear thermal expansion generates polymerization shrinkage which creased internal stresses and causes rupture at the bond interface. This is one of the main causes of gap formation at the tooth-resin interface [5-8]. The type of failure that may occur between the hybrid layer and the tooth structure causes considerable clinical consequences, such as bacterial invasion, recurrent caries development, dentinal sensitivity and pulp irritation [9, 10].

Over the last few decades, the search for a polymer material with a capacity to bond to dental structures has become one of the main study projects [11]. Various load particles have been added to dental adhesives to strengthen the link of bond strength to the dental structure, diminishing polymerization shrinking and increasing the modulus of elasticity of the adhesive layer [12]. Recently, load nanoparticles [13-16], especially nanoparticles of ceramic materials [15, 16], have been used in the formulation of composites, with the purpose of improving physical and mechanical properties.

Of the many synthetic materials studied as substrate support for the development of new materials, hydroxyapatite $[Ca_{10}(PO_4)_6(OH)_2]$ has been widely used and considered one of the most biocompatible ceramics, due to the similarity to the mineral constituents of human bones and teeth [17]. Hydroxyapatite nanoparticles have been used to increase the mechanical properties of resin composites, observing the bioactivity and bond of these composites to the dental structure [18]. In reports in the literature [19] the incorporation of hydroxyapatite in polymer adhesives has also significantly improved the degree of monomer conversion and polymerization rate. Thus, hydroxyapatite nanoparticles may be a promising option of material for the preparation of new dental adhesives with superior properties. The aim of this study was to modify a polymer dental adhesive system with hydroxyapatite nanoparticles as load, characterize it by means of light microscopy, scanning electron microscopy, Energy dispersive X-ray spectroscopy, X-ray diffraction and Fourier Transformed infrared spectroscopy technique (ATR-FTIR) and evaluate the effectiveness of its bond to dental structure.

EXPERIMENTAL PROCEDURE

The study was submitted to the Research Ethics Committee of the Federal University of Vales do Jequitinhonha e Mucuri, under protocol No.186/10.

Preparation of Experimental Units

The sample size calculation was made from a pilot study, considering the degree of confidence of the sample of 95% and statistical power of 95% desired and estimated. The tolerable error of sampling was 4% [20], stipulating a sample size of fifteen.

Sixty-six human pre-molar teeth, whole and recently extracted due to orthodontic indication in subjects under the age of 21 years were selected, well cleaned and autoclave sterilized (Cristófoli Vitale, Cristófoli, Brazil) at 120°C, working pressure of 1 kgf/cm^2, for 20 min [21]. Each of the teeth was section in its horizontal direction, using a diamond disc in a metallographic cutter (ELSAW, Elquip, Brazil), thus obtaining the middle third of the tooth crown. Two sections for test were obtained by sectioning each part of the tooth crown in its vertical direction (Figure 1(a)). The sections were embedded in a chlorinated polyvinyl chloride tube (CPVC), (AMANCO, Brazil) 2 cm in diameter and 1.5 cm high, and filled with polyester resin. In half of the embedded sections, enamel surfaces were exposed (Figure 1(b)) and in the other half, dentin surfaces (Figure 1(c)). Initially, the exposed surfaces were polished with 400 grain abrasive paper in a polishing machine and metallographic grinder (PLF-DV, Fortel, Brazil) with constant cooling. After this,

600 grain abrasive paper was used to simulate the smearlayer. The test specimens (tsps) were washed and stored in distilled water at a temperature of 37°C ± 1°C in an oven (Fanem, Brazil), and were then randomly distributed into ten groups (n = 15), according to Table 1.

With the exception of Groups G5E and G5D, enamel and dentin surface etching of the other groups was performed with 37% phosphoric acid for 30 and 15 s, respectively, and was followed by washing for the same length of time.

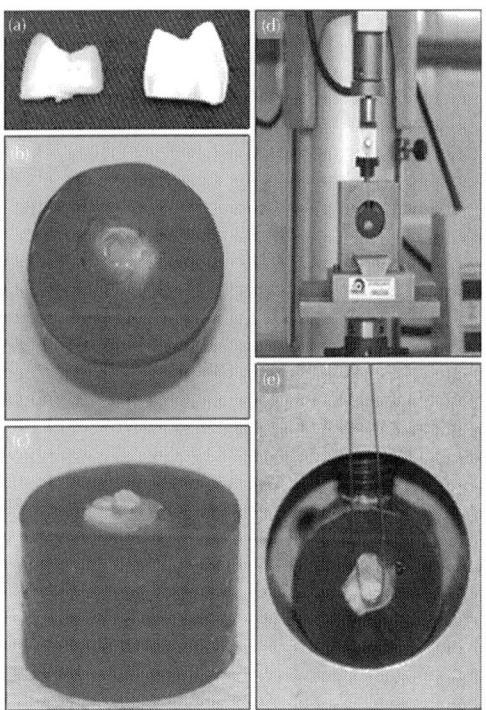

Figure 1: Tooth sections (a) embedded in CPVC, showing exposure of enamel (b) and dentin (c) surfaces, with simulation of resin composite restoration, device used for fixating the tooth-polyester resin sets to the universal test machine for the microtensile test, with orthodontic wire tied to the device and the load cell at the upper extremity of the machine (d) to involve the area closest to the interface of the restoration and dental structure during the test (e).

Table 1: Identification of the study groups and materials used, with their respective commercial brands, manufacturers and organic and inorganic constitution

	EXPERIMENTAL GROUPS (n = 15):				
	G1E* G1D**	G2E G2D	G3E G3D	G4E G4D	G5E G5D
Acid etching	37% Phosphoric acid (3M ESPE)				-
Adhesive System	Adper™ Scotchbond Multi-uso (3M ESPE) Conventional 1-Hydrophilic Monomer 2-Hydrophobic Monomer	Adper™ Scotchbond Multi-uso (3M ESPE) Conventional 1-Hydrophilic Monomer 2-Hydrophobic Monomer + HAP	Adper™ Scotchbond Multi-uso (3M ESPE) Conventional 1-Hydrophilic Monomer + HAP 2-Hydrophobic Monomer	Adper™ Single Bond 2 (3M ESPE) Monocomponent 1-Hydrophilic Monomer + Hydrophobic Monomer	GO! Single Dose (SDI) Self-etching 1-Acid + Hydrophilic Monomer + Hydrophobic Monomer
	Organic Matrix 1-HEMA, polyalkenoic acid Alcohol and water 2-Bis-GMA, HEMA, Camphorquinone	Organic Matrix 1-HEMA, polyalkenoic acid Alcohol and water 2-Bis-GMA, HEMA, Camphorquinone	Organic Matrix 1-HEMA, polyalkenoic acid Alcohol and water 2-Bis-GMA, HEMA, Camphorquinone	Organic Matrix Bis-GMA, HEMA, dimethacrylate, polyacrylic and poly itaconic acids Alcohol and water	Organic Matrix Does not have HEMA Fluoride
	Inorganic Load -	Inorganic Load HAP Nanoparticles 10% (volume)	Inorganic Load HAP Nanoparticles 10% (volume)	Inorganic Load Silicon 10% (volume)	Inorganic Load (not found)
Resin Composite	Microhybrid Filtek Z350 (3M ESPE) Organic Matrix: Bis-GMA, UDMA, Bis-EMA, TEGDMA, PEGDMA Inorganic Load: Zirconium, Silica - 63.3% (volume)				

*E-Enamel; **D-Dentin.

The surfaces of G1E and G1D were dried with absorbent paper. A thin coat of hydrophilic monomer (Table 1) was applied with a microbrush (Vigodent, Brazil), restrictedly on the etched surfaces, which were softly dried for 5 s to allow complete solvent evaporation. Two consecutive coats of hydrophobic monomer (Table 1) were applied with a microbrush, without excess, and submitted to a light jet of air to favor flow over the dental structure. The hydrophobic monomer was light activated for 10 s. After this a metal matrix (1 mm in diameter and 2 mm deep) was placed over the enamel and dentin surfaces of the tsps, in which restorations were simulated (Figures 1(b) and (c)) with nanoparticulate resin composite FILTEK Z350 (3M Espe, Brazil) (Table 1) and light activated for 40 s.

The nanoparticles were obtained by means of hydroxyapatite synthesis by the wet process based on the precipitation route, with calcium hydroxide and orthophosphoric acid as precursors, with a molar Ca/P ratio fixed at 1.67 [22]. The HAP nanoparticles were weighed on an analytical scale (Al 204, Mettler Toledo, Brazil) and mixed with the hydrophobic and hydrophilic monomers at a concentration of 0.10 g/ml, using a magnetic agitator (ARE, Velp Scientifica, Brazil), at a speed of 360 rpm for 30 s.

In Groups G2E and G2D and in G3E and G3D the same procedures were performed as for the previously mentioned groups, however, using the hydrophobic and hydrophilic monomers, modified with HAP, respectively. In G4E and G4D two consecutive coats of monocomponent adhesive system (Table 1) were applied with a microbrush on the etched enamel and dentin surfaces. The adhesive system was gently dried for 5 s and light activated for 10 s.

On the etched surfaces of G5E and G5D the excess water was removed with absorbent paper. One coat of self-etching adhesive system (Table 1) was applied with a microbrush, restrictedly on the surfaces, actively for 20 s. After this, the adhesive was dried with a jet of air at high pressure for 5 s, for complete solvent evaporation. The adhesive was light activated for 10 s.

Nanoparticulate composite resin Z350 was used to perform the restorations in all the other groups, in the same way as performed in G1E and G1D. The materials were used strictly in accordance with all the respective manufacturers' instructions. The materials were light activated using a light emitting diode appliance (LED) Optilight LD Max (Gnatus, Brazil) at the interval of 800 mW/cm², monitored by a radiometer.

The tsps were stored in distilled water at 37°C (±1°C) for 14 days.

Bond Strength Test

After drying at ambient temperature for 12 h, the tsps were submitted to the microshear bond strength test in a Universal test machine EZ Test (Shimadzu, Japan) with a load cell of 200 Kgf at a speed of 0.5 mm/min to measure the bond strength of the materials. For this test, a circular metal device was used to hold the specimen and fixate the set. An orthodontic wire 2 mm thick was tied to a fixed base on the load cell placed at the top part of the test machine (Figure 1(d)), involving the area closest to the base of the restoration (Figure 1(e)), which was subjected to traction until rupture. The bond strength was calculated automatically by the software program (Trapezium, version 1.1.5, Shimadzu, Japan) of the equipment, which is generally measured with the apparent failure load divided by the surface area. The results obtained were transformed into MPa.

Fracture Pattern Analysis

After analysis of the fracture areas of all the tsps, by means of light microscopy (LM) under a stereomicroscope (Stemi 2000C, Carl Zeiss, Canada), at 15X magnification, the type of failure that occurred at the tooth/ adhesive system/restoration interface was classified according to Fowler [23], with modifications: 1 - adhesive, fracture of the adhesive system, hydrophobic monomer present in the restorative resin composite or in the tooth structure or in both; 2 - cohesive in resin, fracture in the restorative resin composite, resin present on both sides of the test specimen; 3 - cohesive on dental structure, fracture in enamel or dentin, dental tissue present on one of the two remaining sides of the test specimen; 4 - mixed, presence of two or more types of fractures described above.

Characterization of Materials

Scanning Electron Microscopy and X-Ray Dispersive Energy Spectroscopy

The fracture interfaces representative of each group were sputter-coated with goldpalladium, and analyzed by scanning electron microscopy (SEM), in equipment CS- 3500 (Shimadzu, Japan). Qualitative element analysis of the tsps was performed by energy dispersive X-Ray spectroscopy (EDX), in a spectrophotometer CS3200 (Oxford, England).

X-Ray Diffraction

The composites developed were submitted to X-ray diffraction (DRX) in a diffractometer RXD6000 (Shimadzu, Japan), with monochromatic K Cu radiation (1.5406 A°) and operational tube with a voltage of 40 kV and current of 30 mA.

Fourier Transform Infrared Spectroscopy

The samples of the composites developed were analyzed by means of Fourier Transform Infrared Spectroscopy (FTIR), by the attenuated total reflectance technique (ATR), (Nicolet 6700, Thermo Electron, USA). Spectra were obtained with 32 scans in the interval between 675 and 4000 cm^{-1}, with a resolution of 4 cm^{-1} and units of absorbance (abs).

Statistical Analysis

After performing the normality test (Shapiro-Wilk), a non-parametric statistical test (Kruskal-Wallis) was applied to verify differences between the groups, And after this, the Mann-Whitney test to verify the intragroup differences, using the software program Statistical Package for Social Sciences (SPSS for Windows, version 17.0, SPSS Inc., USA). To evaluate the type of failure that occurred between the different groups, descriptive data analysis was applied, represented by means of graphs.

RESULTS

The statistical results showed that there was significant difference between the groups ($p < 0.05$), as regards bond strength after mechanical testing in enamel and dentin (Table 2). After microanalysis under stereomicroscope, adhesive fractures were observed in the majority of the specimens on both tooth surfaces. A larger number of cohesive fractures in enamel were observed in the groups with the adhesive systems modified with HAP (G2E and G3E). In the monocomponent adhesive system (G4E), in addition to the presence of cohesive fractures in enamel, mixed fractures were also observed. Cohesive fractures in the dentinal structure were observed only in Group G3D (Figure 2).

Table 2: Values of bond strength to enamel and dentin, in the groups of materials studied, and respective statistical analyses

Group n = 15	Bond Strength (MPa): (SD) *p < 0.001	**Mann-Whitney
G1E	50.51 (8.04)	A
G2E	60.69 (15.29)	AB
G3E	64.40 (7.36)	B
G4E	63.28 (20.29)	BC
G5E	7.57 (3.32)	D
G1D	23.89 (16.44)	AC
G2D	32.99 (16.06)	AB
G3D	39.59 (21.46)	B
G4D	13.27 (11.30)	CD
G5D	6.60 (2.77)	D

*Kruskal-Wallis. **Equal letters indicate there is no statistically significant difference between the groups.

The SEM micrographs of samples representative of the specimens analyzed by LM showed a thin layer of adhesive on the enamel (Figure 3(a)) and dentin surfaces (Figure 3(b)) in the majority of fracture areas of the dental remainders, suggesting the adhesive type of fracture. Spectra of EDS of these surfaces in enamel (Figure 3(a)) and dentin surfaces (Figure 3(b)) showed high peaks of Calcium

(Ca) and Phosphorous (P) corresponding to the dental structure, and peaks of low intensity of Silicon (Si) corresponding to the inorganic load of the monocomponent adhesive system. Cohesive fractures in enamel (Figure 3(c)), dentin (Figure 3(d)), restorative resin (Figure 3(e)) and mixed fractures (Figure 3(f)) were also observed at the interfaces. The spectra of EDS of cohesive fractures in enamel (Figure 3(c)), and in dentin (Figure 3(d)) presented peaks of high intensity of Ca and P, while the spectra of EDS of cohesive fracture in the restorative resin (Figure 3(e)) showed peaks of high intensity of Si and Zirconia (Zr) and low intensity of Ca. Mixed fracture in enamel (Figure 3(f)) presented spectra with peaks of Ca, P, Zr and Si. Peaks of Gold and Palladium from the sputter-coating of the materials could also be visualized; however, they were not identified in the spectra.

The crystallography of the HAP nanoparticles was analyzed by the spectra generated by DRX. The information obtained was compared with standardized records from the database of the International Center for Diffraction (JCPDS—International Centre for Diffraction Data, USA) for the calcium phosphate materials, and it was possible to identify the material as being hydroxyapatite. In the qualitative DRX analysis of the HAP nanoparticles, a high degree of crystallinity was observed. In the diffractograms in Figure 4(a), the peaks of higher intensity of the $Ca_{10}(PO_4)_6(OH)_2$ phase, characteristic of HAP may be visualized, and the presence of a large quantity of amorphous phase on the adhesive of the conventional adhesive system (G1) and in the adhesive with the addition of HAP (G2). In the diffractograms in Figure 4(b) HAP is shown with the presence of crystalline peaks, hydrophobic monomer of the conventional adhesive system with characteristics of amorphous material (G1) and the hydrophilic monomer modified with HAP (G2) with an amorphous phase and peaks characteristic of HAP. A small displacement of HAP peaks (a) may be observed in the diffractograms of Group G3 (Figure 4(b)). In the FTIR spectra (Figure 5) it was observed that the absorption bands with reference to

the PO_4^{-3} ions of HAP appeared in the spectra in the vibrational mode of stretching (v_3) and were evident at 1088, 1039 and 1032 cm^{-1}; the band with reference to CO_3^{-2} was shown at 864 cm^{-1}. The presence of a narrow band at 3594 cm^{-1} was observed in the region corresponding to the vibrations of stretching of structural OH-ions. The FTIR spectra of the hydrophobic monomer of Group G1 presented absorption bands with reference to Bis-GMA at 1250, 1297, 1442, 1507, 1635, 1721 and 2962 cm^{-1} and HEMA at 1639 cm^{-1} corresponding to the methacrylate group (C=C), 1721 cm^{-1} to C=O and 1182 cm^{-1} to C-O. The spectra of the hydrophobic monomer modified with HAP (G2) showed absorption bands similar to those of the adhesive in G1, with little intensity of absorption of the phosphate bands of HAP. The hydrophilic monomers of G1, which contain HEMA, presented the same bands as previously described. In G3, the hydrophilic monomer modified with HAP showed evident bands of HAP, free OH groups at 3243 cm^{-1} and wide absorption bands from 3374 cm^{-1} corresponding to the OH groups of HAP and HEMA.

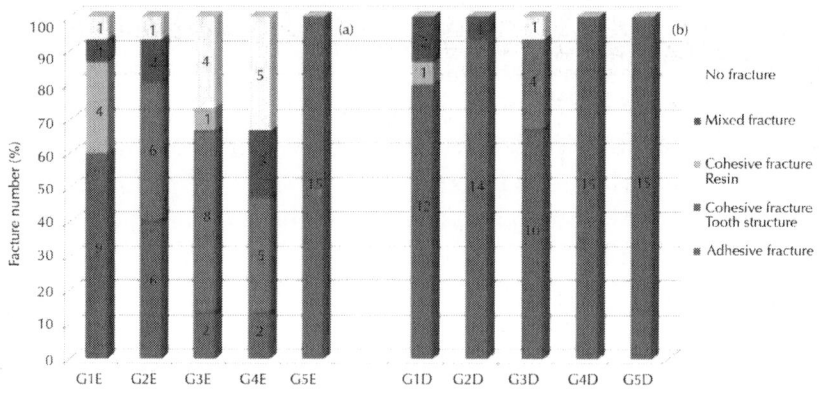

Figure 2: Graphic representation of the number of fractures (%) of the specimens in the Enamel (E) and Dentin (D) groups, after microanalysis by LM.

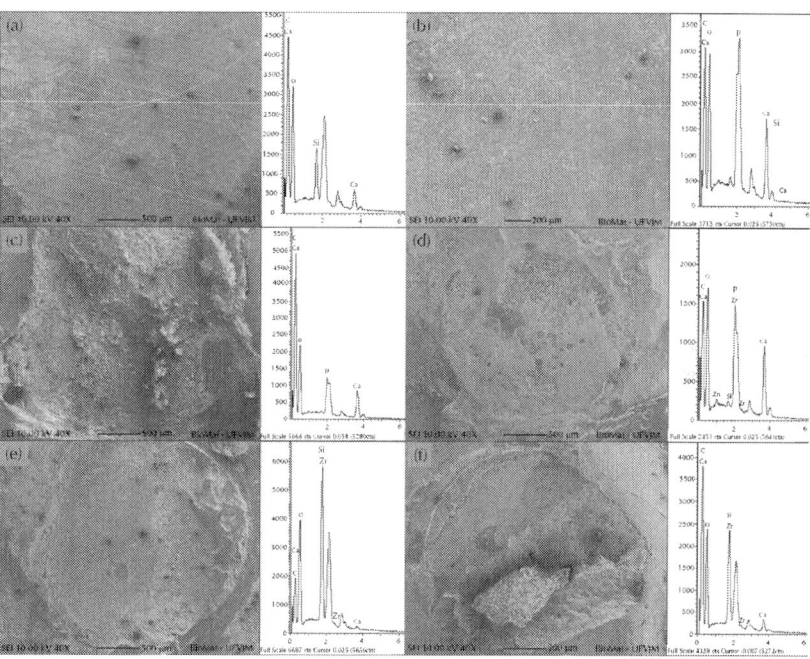

Figure 3: SEM Micrographs and respective EDX spectra of samples representative of the experimental groups, showing adhesive fracture areas in enamel: G2 (a) and in dentin: G3 (b); cohesive fracture areas in enamel: G2 (c) and in dentin: G3 (b); cohesive fracture areas in resin: G1 (e); and mixed fracture: G4 (f).

DISCUSSION

Premolar teeth extracted from individuals under the age of 21 years were used with the purpose of avoiding the interference of variables in the bond strength values. Physiological alterations resulting from aging of the dentinal tissue increase the degree of dentin mineralization, increasing its thickness and reducing its permeability [24].

The insertion of inorganic particles promoted improvements in the mechanical properties of the polymers, considering that the composites modified with HAP nanoparticles (G2 and G3) presented the highest bond strength values in comparison with the precursor material G1, to both dental surfaces. As the volumetric fraction of the inorganic matrix increases, the link of bond force is strengthened, because the diminishment of the organic portion minimize the polymerization shrinkage and increases the modulus of elasticity of the adhesive layer [11,25].

The lower bond strength values found in G2E and G2D in comparison with G3E and G3D, respectively, may be associated with the low conversion of monomers during the light activation process; that is to say, it is determined by the percentage of double carbon bonds (C=O) which were broken and transformed into simple bonds [26]. The result of this process is directly connected to the behavior of adhesive bonds to dental substrate [26]. The incorporation of HAP nanoparticles into the hydrophobic monomer may have made the interaction of these particles difficult by means of dipole-dipole type bonds to the dental structure, and also enabled a greater formation of HAP agglomerates. These, in turn, may have altered the rate of refraction of the wave bands of the light activator appliances to the adjacent layers, altering the degree of polymerization and diminishing the formation of cross links of these polymer systems, generating polymers with a larger quantity of free monomers, and being more susceptible to degradation. Other significant factors were the possibility of agglomerates of HAP nanoparticles having obliterated the enamel prisms and dentinal tubules, difficulty of flowing of the adhesive modified with HAP, and the reduction in the formation of resin tags, responsible for mechanical retention.

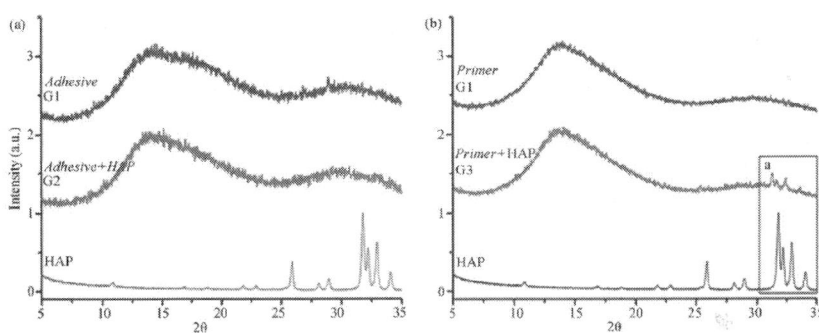

Figure 4: Diffractograms with peaks of greater intensity of HAP, of the hydrophobic monomer of the conventional adhesive system (G1) and of the hydrophobic monomer modified with HAP nanoparticles (G2) (a); diffractograms with peaks of greater intensity of HAP, of the hydrophilic monomer of the conventional adhesive system (G1) and of the hydrophilic monomer modified with HAP nanoparticles (G3) (b), showing peaks characteristic of HAP (a).

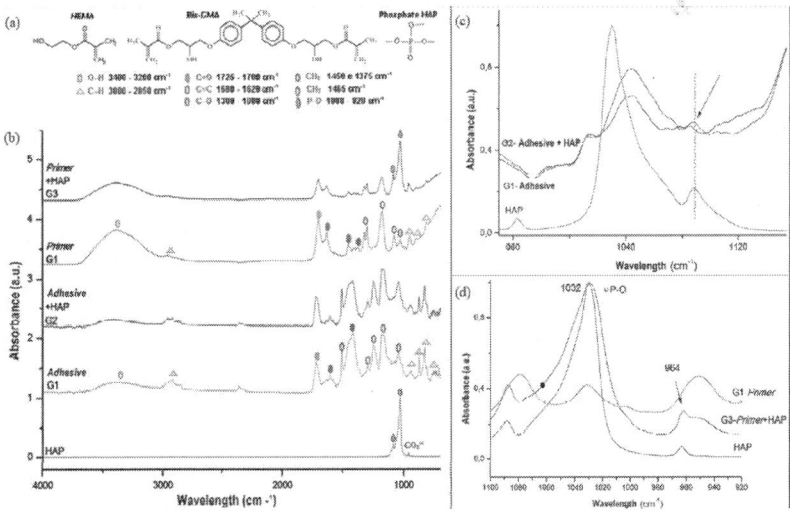

Figure 5: Functional Groups of HEMA and Bis-GMA molecules and of the phosphate group of HAP (a); FTIR spectra with the absorption bands in the infrared with reference to the chemical bonds of the functional

groups of the composition of HAP, of the hydrophobic monomer of the conventional adhesive (G1), of the hydrophobic monomer modified with HAP nanoparticles (G2), of the hydrophilic monomer of the conventional adhesive system (G1) and of the hydrophilic monomer modified with HAP nanoparticles (G3) (b); FTIR spectra with the absorption bands in the infrared with references to the chemical bonds of the functional groups of the composition of HAP, of the hydrophobic monomer of the adhesive (G1) and of the hydrophobic monomer of the adhesive modified with HAP nanoparticles (G2) (c); FTIR spectra with absorption bands in the infrared with reference to chemical bonds of the functional groups of the composition of HAP, of the hydrophilic monomer of the adhesive (G1) and of the hydrophilic monomer of the adhesive modified with HAP nanoparticles (G3) (d).

The hydrophilic monomer may have acted as a lining of the inorganic load, enabling an improvement in the bond at the HAP nanoparticle interface with the resin matrix of the hydrophobic monomer [27]. The sealing of the HAP surface increases its dispersion and minimizes the formation of inorganic aggregates [27], also making it possible for greater cohesion of these particles with the dental structure to occur.

The lowest bond strength value was found in G4D in comparison with G1D, G2D and G3D, while G4E presented the second highest bond strength value. One of the explanations for this occurrence is the greater complexity of the adhesive bond to dentin [28]. Another probable explanation is related to demineralized dentin drying. The collapse of collagen fibrils and establishment of hydrogen bridges between them may occur, contracting and tensing the collagen network, and leading to the loss of the tridimensional spatial configuration. Thus, thus the permeability of this network is diminished, which hinders the infiltration of the adhesive monomers, leading to the formation of a thinner hybrid layer with deficient sealing of the dentinal tubules [29, 30].

The monomer/solvent combination in adhesive systems with their application performed in a larger number of steps must be capable of breaking the interpeptide hydrogen bridges, and thus re-expand the dentin matrix, re-creating the interfibrillar spaces of the collapsed collagen network to allow infiltration of the hydrophobic monomer [29, 30].

The bond strength, measured by means of the microshear test, varied with the treatment of the exposed tooth surface. Inferior behavior was observed for Groups G5E and G5D, in which the self-etching adhesive system was used, and which removes a micrometric layer of dental surface [31-33]. Greater unobstruction of enamel and dentinal tubule prisms was verified with the phosphoric acid etching process, [34] as was used in the other experimental groups, allowing better flow of the hydrophobic monomer, with more voluminous and deeper tag formation [32-34].

Analysis of the fracture surfaces of the tsps by LM identified a larger number of adhesive failures, which may be justified by the better stress distribution at the bond interface of small specimens demanded in the microshear test [34]. In Groups G2E, G3E and G4E a larger number of cohesive fractures in the dental structure were observed. This is justified by the higher bond strength values in the mentioned groups, with emphasis on G2E and G3E, in which the HAP nanoparticles may have interacted with the hydroxyapatite of enamel tissue, strengthening the bond even further.

In Groups G3E and G4E there were a significant number of specimens whose bond strength value was higher than the rupture strength of the orthodontic wire used in the microshear test, preventing their fracture. This may be associated with the presence of HAP nanoparticles in G3E and silicon in G4E, promoting greater bond strength.

The capacity of HAP nanoparticles to bond to the dentinal structure is potentiated when mixed with the hydrophilic monomer, and this may justify the occurrence of cohesive fractures in dentin in Group G3D. The relationship between carbonate and apatite is of great importance, since carbonate increases the chemical reactivity of apatite, particularly by the increase in the product of solubility and dissolution rate [35].

The clinical consequences of failures that may occur at the tooth/adhesive system interface depend on where they occur at the bond interface. Only if the failure occurs between the base of the hybrid layer and the subjacent dentin can it enable the demineralization of the sealed dentin, leading to bacterial invasion, dentinal sensitivity and pulp irritation [10].

The microshear test was used in this study in accordance with the specifications of the ISO 527-1 and 527-2 standards for determining the properties of materials [36] Imperfections or defects at the bond interfaces, such as air bubbles, water or solvent may allow the development of local stresses during the bond strength test in larger areas. The gaps formed may propagate rapidly and cause failures. In smaller areas, there are fewer imperfections, bearing higher pressures before fracturing [37]. The speed of 0.5 mm per minute used in the test was also an important factor for a uniform distribution of the force. The ISO standard recommends a speed of at least 0.45 mm/min to a maximum of 1.05 mm/min [38].

The bonds between dentin and resin are friable, and tend to fail when subjected to various types of stresses, such as those that typically occur in the mouth. It is common for standard deviations of 30% to 40% to occur around the mean values, as mentioned in TC106 ISO (TR 11405) for the measurement of bond strength to the dental substrate [38].

Element analysis by EDX of the adhesive fracture areas on the dental remainder identified not only the Si in the organic matrix of the hydrophobic monomer, but also Ca and P, probably from the tooth structure located below this thin layer. A large quantity of Ca and P was observed at the interface of fractures of the cohesive type in dental structure, chemical elements of dental tissue, in addition to peaks of Si, characteristic remnants of the organic matrix of the hydrophobic monomers used. In the cohesive fractures in resin, in addition to Ca and P, peaks of Zr and Si - components of the restorative resin composite were found. Mixed fractures are well characterized by the presence of spectra with peaks of Ca and P and of Zr and Si, characteristic elements of the hydrophobic monomer and restorative resin composite. In the cohesive fracture areas in enamel and dentin of Group G3, we observed spectra with higher peaks of Ca and P coming from the dental structure and

HAP nanoparticles incorporated into the hydrophilic monomer of the adhesive system.

Qualitative analysis by X-ray diffraction is a powerful method of chemical analysis, considering the entire crystalline structure has a diffraction pattern and that the same substance always presents the same model [39]. Thus, in the analysis of the diffractograms of the studied materials, it was observed that HAP presented standardized peaks, in agreement with reports in the literature [20, 40, and 41]. The presence of a large quantity of amorphous phase in the hydrophobic monomer of Group G1 is owing to the presence of HEMA and Bis-GMA as constituents of its organic matrix. The presence of a large halo of the amorphous phase in the diffractograms, probably supplied the standard peaks in the DRX of HAP, which was added to the material as inorganic load in the proportion of only 10% of its total load. This small quantity of HAP was already sufficient to increase the viscosity of the hydrophobic monomer even further, which contains Bis-GMA, a component considered viscous due to the intermolecular union of hydrogen, by the presence of two methacrylate groups in its molecule, as well as aromatic rings and hydroxyl groups [8]. The formation of agglomerates of HAP nanoparticles in the hydrophobic monomer diminished its flowability and penetration into the etched enamel and dentin surfaces, and consequently, made it difficult for them to bond.

The hydrophilic monomer in Group G1 had only HEMA in its composition. This less viscous monomer with the characteristic of hydrophilicity is considered a wetting agent which, associated with polyalkenoic acid and also to alcohol and water as solvent, facilitates the penetration of the adhesive into the dental structure [8]. These agents would probably allow the dispersion of the HAP nanoparticles, which made it possible to visualize the HAP peaks of low intensity in the diffractograms of G3. No formation of a new phase detectable by the characterization method used was observed. In the diffractograms of the hydrophobic and hydrophilic monomers modified with HAP there were absorption bands characteristic of HAP; a contribution of the HAP nanoparticles to thickening the absorption band of these modified composites was also observed, allowing one to suppose the occurrence of strong interatomic bonds of the functional and specific groups present in the constitution of materials [39].

The molecular structure of HAP and the composites developed presented vibrational excitations resulting from the strong interatomic bonds (covalent bonds) of the specific functional groups present in the materials. The graphs resulting from the energy absorbed by the sample, by reason of the wave number in which this energy was absorbed are represented by the FTIR spectrum [42]. The profile of the FTIR spectrum of HAP showed agreement with the spectra found in the literature [20, 40, and 41]. In this FTIR spectrum of hydrophobic monomers modified with HAP (G2) no evident modifications were observed from the inclusion of HAP as its inorganic constituent, showing it to have the same characteristics as the bands presented by the presence of Bis-GMA [41] in the hydrophobic monomer. To the contrary, the hydrophilic monomer modified with HAP (G3) presented absorption bands of HEMA [44] end more intense absorption bands at 830 cm^{-1}, 1044 cm^{-1} and 1090 cm^{-1} belonging to the phosphate groups of HAP. This fact added to the presence of bands that identify free OH groups, probably contributed to the higher bond strength of this composite to both enamel and dentin.

The conventional adhesive system without load is a low cost material, available on the dental market, which modified with HAP nanoparticles processed with accessible reagents and by simple techniques, made it possible for a true bond to the dentin substrate to occur. This study enables a review of scientific knowledge and development of polymer composites with more effective bonds to enamel, and particularly to dentin. In addition, it incites new methodologies to be performed to confirm or refute the results found.

CONCLUSIONS

Considering the experimental conditions of this study it could be concluded that:

- The addition of HAP nanoparticles to the hydrophilic monomer of a conventional adhesive system increased the bond strength to enamel and dentin;
- The bond strength to the dental structure, of a HAPmodified hydrophobic monomer of the conventional adhesive system presented limited improvement;
- The bond strength to the dental structure, of the hydrophilic monomer with HAP, of the conventional adhesive system was shown to be higher than that of the monocomponent and self-etching adhesive systems;
- The majority of the fractures that occurred at the adhesive system/dental structure interfaces of the composites were of the adhesive type, and the groups with the presence of HAP in their constitution presented a higher percentage of cohesive fractures in enamel;
- The composites developed were shown to be stable, with the presence of crystallographic and amorphous phases, in addition to a similar morphology and chemical structure to those of their precursor materials.

ACKNOWLEDGEMENTS

The authors thank the "Laboratório Multiusuário de Microscopia Avançada—LMMA" (FAPEMIG-Processo CEX 112/10) of the Department of Chemistry/UFVJM for the obtainment of the XRD spectra; the Pathology Laboratory of the Department of Basic Sciences/UFVJM, and thank CAPES, SECTES/FAPEMIG and CNPq for the financial support.

REFERENCES

1. A. B. Matos, C. H. C. Saraceni, M. M. Jacobs and M. Oda, "Estudo de Resistência à Tração de Três Sistemas Adesivos Associados a Resina Composta em Superfícies Dentinárias," Pesquisa Odontológica Brasileira, Vol. 15, No. 2, 2001, pp. 161-165.http://dx.doi.org/10.1590/S1517-74912001000200014

2. K. L. Van Landuyt, J. Snauwaert, J. De Munck, M. Peumans, Y. Yoshida, A. Poitevin, E. Coutinho, K. Suzuki, P. Lambrechts and B. Van Meerbeeka, "Systematic Review of the Chemical Composition of Contemporary Dental Adhesives," Biomateriais, Vol. 28, No. 26, 2007, pp. 3757- 3785. http://dx.doi.org/10.1016/j.biomaterials.2007.04.044

3. G. Inoue, T. Nikaido, M. R. Foxton and J. Tagami, "The Acid-Base Resistant Zone in Three Dentin Bonding Systems," Dental Materials Journal, Vol. 28, No. 6, 2009, pp. 717-721. http://dx.doi.org/10.4012/dmj.28.717

4. M. Sarr, A.W. Kane, J. Vreven, A. Mina, K.L. Van Landuyt, M. Peumans, P. Lambrechts, B. Van Meerbeek and J. De Munck, "Microtensile Bond Strength and Interfacial Characterization of 11 Contemporary Adhesives Bonded to Bur-Cut Dentin," Operative Dentistry, Vol. 35, No. 1, 2010, pp. 94-104. http://dx.doi.org/10.2341/09-076-L

5. T. Yoshikawa, N. Wattanawongpitak, E. Cho and J. Tagami, "Effect of Remaining Dentin Thickness on Bond Strength of Various Adhesive Systems to Dentin," Dental Materials Journal, Vol. 31, No. 6, 2012, pp. 1033-1038. http://dx.doi.org/10.4012/dmj.2012-143

6. N. B. Suryakumari, P. S. Reddy, L. R. Surender and R. Kiran, "In Vitro Evaluation of Influence of Salivary Contamination on the Dentin Bond Strength of One-Bottle Adhesive Systems," Contemporary Clinical Dentistry, Vol. 2, No. 3, 2011, pp. 160-164.http://dx.doi.org/10.4103/0976-237X.86440

7. R. N. Garcia, A. E .G. Alvarez, C. E. Dias, M. A. Mazaro, T. Firmo, H. Stuker and M. Giannini, "Bond Strength of Contemporary Restorative Systems to Enamel and Dentin," Revista Sul-Brasileira de Odontologia, Vol. 8, 2011, pp. 60-67.

8. A. Peutzfeldt, "Resin Composites in Dentistry: The Monomer Systems," European Journal of Oral Science, Vol. 105, No. 2, 1997, pp. 97-116. http://dx.doi.org/10.1111/j.1600-0722.1997.tb00188.x

9. L. M. Cavalcante, L. F. J. Schneide, L. S. Silva, A. K. Bedran-Russo and L. A. F. Pimenta, "Efeito da Ciclagem Térmica na Microinfiltração e Microtração de Restaura- **ções** de Resina Composta," Revista da Faculdade de Odontologia-UPF, Vol. 14, 2009, pp. 132-138.

10. J. Perdigão, "Dentin Bonding—Variables Related to the Clinical Situation and the Substrate Treatment," Dental Material, Vol. 26, No. 2, 2010, pp. 24-37.http://dx.doi.org/10.1016/j.dental.2009.11.149

11. M. Sadat-shojai, M. Atai, A. Nodehi and L. N. Khanlar, "Hydroxyapatite Nanorods as Novel Fillers for Improving the Properties of Dental Adhesives: Synthesis and Application," Dental Materials, Vol. 26, No. 5, 2010, pp 471- 482.http://dx.doi.org/10.1016/j.dental.2010.01.005

12. U. Lohbauer, A. Wagner, R. Belli, C. Stoetzel, A. Hilpert, H. D. Kurland, J. Grabow and F. A. Müller, "Zirconia Nanoparticles Prepared by Laser Vaporization as Fillers for Dental Adhesives," Acta Biomaterialia, Vol. 6, No. 12, 2010, pp. 4539-4546. http://dx.doi.org/10.1016/j.actbio.2010.07.002

13. M. A. Ribeiro, I. X. Ferreira, R. P. Lima, A. L. A. Mariz, J. G. F. Pompeu and C. H. V. Silva, "The Insertion Technique's Infl Uence of Composite resin on the Microleakage in Occlusal Esthetic Restorations," Revista Odontológica Clínico Científica, Vol. 9, 2010, pp. 345-348.

14. H. Zhang and B. W. Darvell, "Mechanical Properties of Hydroxyapatite Whisker-Reinforced Bis-GMA-Based Resin

Composites," Dental materials, Vol. 28, No. 8, 2012, pp. 824-830.http://dx.doi.org/10.1016/j.dental.2012.04.030

15. V. C. Leitune, F. M. Collares, R. M. Trommer, D. G. Andrioli, C. P. Bergmann and S. M. Samuel, "The Addition of Nanostructure Hydroxyapatite to an Experimental Adhesive resin," Journal of Dentistry, Vol. 41, No. 4, 2013, pp. 321-327.http://dx.doi.org/10.1016/j.jdent.2013.01.001

16. A. Akhavan, A. Sodagar, F. Motjahedzadeh and K. Sodagar, "Investigating the Effect of Incorporating nanosilver/ Nanohydroxyapatite Particles on the Shear Bond Strength of Orthodontic Adhesives," Acta Odontologica Scandinavica, Vol. 71, No. 5, 2013, pp. 1038-1042. http://dx.doi.org/10.31 09/00016357.2012.741699

17. J. Vandiver, D. Dean, N. Patel, W. Bonfield and C. Ortiz, "Nanoscale Variation in suRface Charge of Synthetic Hydroxyapatite Detected by Chemically and Spatially Specific High-Resolution Force Spectroscopy," Biomaterials, Vol. 26, No. 3, 2005, pp. 271-283.http://dx.doi.org/10.1016/j. biomaterials.2004.02.053

18. A. S. Khan, K. R. Hassan, S. F. Bukhari, W. Ferranti and S. L. Rehmaniu, "Structural and in Vitro Adhesion Analysis of a Novel Covalently Coupled Bioactive Composite," Journal of Biomedical Materials Research Part B, Vol. 100, No. 1, 2012, pp. 239-248.http://dx.doi.org/10.1002/jbm.b.31945

19. Y. Zhang and Y. Wang, "Hydroxyapatite Effect on Photopolymerization of Self Etching Adhesives with Different Aggressiveness," Journal of Dentistry, Vol. 40, No. 7, 2012, pp. 564-570. http://dx.doi.org/10.1016/j.jdent.2012.03.005

20. J. K. Jekel, D. L. Katz and J. G. Elmore, "Epidemiology, Biostatistics, and Preventive Medicine," Elsevier, Philadelphia, 2001.

21. P. Jacques and J. Hebling, "Influence of Autoclave Sterilization of Human Teeth on Dentin Bonding," Pesquisa Brasileira em Odontopediatria e Clínica Integrada, Vol. 6, 2006, pp. 9-13.

22. M. H. Santos, M. Oliveira, L. F. P. Souza, H. S. Mansur and W. L. Vasconcelos, "Synthesis Control and Characterization of Hydroxyapatite Prepared by Wet Precipitation Process," Materials Research, Vol. 7, No. 4, 2004, pp. 1- 6. http://dx.doi.org/10.1590/S1516-14392004000400017

23. C. S. Fowler, M. L. Swartz, B. K. Moore and B. F. Rhodes, "Influence of Selected Variables on Adhesion Testing," Dental Materials, Vol. 8, No. 4, 1992, pp. 265-269.http://dx.doi.org/10.1016/0109-5641(92)90097-V

24. J. Perdigão, "Effect of Substrate Age and Adhesive Composition on Dentin Bonding," Operative Dentistry, Vol. 38, 2012.

25. Y. Liu, Y. Tan, T. Lei, Q. Xiang, Y. Han and B. Huang, "Effect of Porous Glass-Ceramic Fillers on Mechanical Properties of Light-Cured Dental Resin Composites," Dental Materials, Vol. 25, No. 6, 2009, pp. 709-715.http://dx.doi.org/10.1016/j.dental.2008.10.013

26. T. F. Boing, G. M. Gomes, C. Z. Grande, A. Reis, J. C. Gomes and O. M. M. Gomes, "Evaluation of the Degree of Conversion of a Composite Resin Using Different Surface Treatments before Final Curing," Revista Dentística on line, Vol. 10, 2011.

27. H. Zhang and M. Zhang, "Effect of Surface Treatment of Hydroxyapatite Whiskers on the Mechanical Properties of Bis-GMA-Based Composites," Biomedical Materials, Vol. 5, 2010.

28. B. Poptani, K. S. Gohil, J. Ganjiwale and M. Shukla, "Microtensile Dentin Bond Strength of Fifth with Five Seventh-Generation Dentin Bonding Agents after Thermocycling: An in Vitro Study," Contemporary Clinical Dentistry, Vol. 3, No. 6, 2012, pp. 167-171.http://dx.doi.org/10.4103/0976-237X.101079

29. A. O. Spazzin, B. Carlini Júnior, R. R. Moraes and M. F. Mesquita, "Bonding to Wet and Dry Dentin: Microtensile Bond Strength and Marginal Leakage," Revista de Odontologia da UNESP, Vol. 37, 2008, pp. 91-96.

30.   T. K. Vaidyanathan and J. Vaidyanathan, "Recent Advances in the Theory and Mechanism of Adhesive Resin Bonding to Dentin: A Critical Review," Journal of Biomedical Materials Research Part B: Applied Biomaterials, Vol. 88, No. 2, 2009, pp. 558-578.http://dx.doi.org/10.1002/jbm.b.31253

31.   P. Senawongse, C. Harnirattisai, Y. Shimada and J. Tagami, "Effective Bond Strength of Current Adhesive Systems on Deciduous and Permanent Dentin," Operative Dentistry, Vol. 29, 2004, pp. 196-202.

32.   P. Banks and B. Thiruvenkatachari, "Long-Term Clinical Evaluation of Bracket Failure with a Self-Etching Primer: A Randomized Controlled Trial," Journal of Orthodontics, Vol. 34, No. 4, 2007, pp. 243-251. http://dx.doi.org/10.1179/146531207225022293

33.   M. A. Montasser, J. L. Drummond, J. R. Roth, L. AlTurki and C. A. Evans, "Rebonding of Orthodontic Brackets. Part II, an XPS and SEM Study," Angle Orthodontist, Vol. 78, No. 3, 2008, pp. 537-44. http://dx.doi.org/10.2319/022707-102.1

34.   A. Manabe, M. Kanehira, W. J. Finger, H. Hisamitsu and M. Komatsu, "Effects of Opacity and Oxygen Inhibition of Coating Resin Composites on Bond Strength to Enamel," Dental Materials Journal, Vol. 28, No. 5, 2009, pp. 552-557. http://dx.doi.org/10.4012/dmj.28.552

35.   C. Elliot, "Structure and Chemistry of the Apatites and other Calcium Orthophosphates: Studies in Inorganic Chemistry 18," Elsevier Science, 1994.

36.   T. Arao and N. Nakabayashi, "Effect of Miniaturized Dumbbell-Shaped Specimen to Identify Bonding of Resin to Bovine Dentin," Journal Japanese of Dental Materials, Vol. 16, 1997, pp. 175-181.

37.   H. Sano, T. Shono, H. Sonoda, T. Takatsu, B. Ciucchi and R. M. Carvalho, "Relationship between Surface Area for Adhesion and Tensile Bond Strength-Evaluation of a Microtensile Bond Test," Dental Materials, Vol. 10, 1994, pp. 236-240.http://dx.doi.org/10.1016/0109-5641(94)90067-1

38. International Organization for Standardization. ISO/ TC106 (TR 11405): Dental Materials—Testing of Adhesion to Tooth Structure, Geneva, 2003.

39. B. D. Cullity and S. R. Stock, "Elements of X-Ray Diffraction," Prentice Hall, New Jersey, 2001.

40. M. H. Santos and H. S. Mansur, "Low-Cost Processing Technology for the Synthesis of Calcium Phosphates/ Collagen Biocomposites for Potential Bone Tissue Engineering Applications," Materials Research, Vol. 10, 2007, pp. 431-436.http://dx.doi.org/10.1590/S1516-14392007000400018

41. M. H. Santos, L. G. D. Heneine and H. S. Mansur, "Synthesis and Characterization of Calcium Phosphate/Collagen Biocomposites Doped With Zn^{2+}," Materials Science and Engineering C, Vol. 28, 2008, pp. 563-571.http://dx.doi.org/10.1016/j.msec.2007.07.002

42. D. L. Paiva, G. M. Lampman, G. S. Kriz and J. R. Vyvyan, "Introduction to Spectroscopy," Cengage Learning, Estados Unidos, 2009.

Chapter 8

Ductile Fracture Characterization for Medium Carbon Steel Using Continuum Damage Mechanics

Stergios Pericles Tsiloufas and Ronald Lesley Plaut

Department of Metallurgical and Materials Engineering, Escola Politécnica, University of São Paulo, São Paulo, Brazil.

ABSTRACT

This paper presents the ductility characterization for a medium carbon steel, for two microstructural conditions, that has been evaluated using the continuum damage mechanics theory, as proposed by Kachanov and developed by Lemaitre. Tensile tests were carried out using loading-unloading cycles in order to capture

materials will fail when the damage reaches a critical value $D_c<1$, when the effective area can no longer resist the applied load, leading to the formation of a macroscopic crack.

In his model, Lemaitre assumes the hypothesis of strain equivalence, which states that the damaged material will have the same constitutive behavior of the virgin material, replacing the stress tensor s by the effective stress tensor $\tilde{\sigma}$, defined as:

$$\tilde{\sigma} = \frac{\sigma}{1-D}$$
(2)

One important consequence of this assumption is that one can define an effective elastic modulus of a damaged material, giving an indirect way to measure the damage in a solid, by monitoring the evolution of the Young modulus with increasing strain:

$$D = 1 - \frac{\tilde{E}}{E}$$
(3)

where $\tilde{E}$ is the effective elastic modulus and E is the elastic modulus for the undamaged material.

Using as basis the thermodynamic of irreversible processes [19], CDM treats the damage as an internal thermodynamic state variable, and so its evolution can be derived assuming the existence of a potential of dissipation f and an associated variable Y, named damage strain energy release rate and defined as [10]:

$$-Y = \frac{\sigma_{eq}^2}{2E(1-D)^2}\left[\frac{2}{3}(1+v)+3(1-2v)\left(\frac{\sigma_H}{\sigma_{eq}}\right)^2\right]$$
(4)

where $\sigma_{ea} = \sqrt{2/3}\|\sigma^D\|$ is the von Mises equivalent stress, σ^D is the deviatoric stress tensor, v is the Poisson's ratio and $\sigma_H = (1/3)\mathrm{tr}(\sigma)$ is the hydrostatic stress. Further, Lemaitre [10] shows that the damage evolution can be written as:

$$\dot{D} = -\frac{\partial\phi}{\partial Y}(1-D)\dot{p}$$
(5)

with $P = \sqrt{2/3}\|\varepsilon^p\|$ defined as the accumulated plastic strain and ε^p being the plastic strain tensor. The choice of a proper potential of dissipation that can represent experimental results is the core of any CDM model. In Lemaitre and Chaboche's model, the hypothesis of isotropic damage, existence of a strain threshold for damage initiation and linear evolution of the damage with the accumulated plastic strain leads to the following equation for damage evolution:

$$\dot{D} = \begin{cases} 0 & \text{for } p < p_D \\ \dfrac{-Y}{S}\dot{p} & \text{for } p \geq p_D \end{cases}$$

(6)

where P_D is the accumulated plastic strain threshold and S is the damage resistance parameter, which are material dependent properties. For the uniaxial stress state, and assuming that the elastic strain can be neglected in comparison to the total strain, the accumulated plastic strain can be considered equal to the principal strain. The damage increases until it reaches a critical value D_c which can be calculated with the following equation [10]:

$$D_c = \frac{D_{1c}}{R_v}\left[\frac{\sigma_u}{\sigma_{eq}}(1-D)^2\right]$$

(7)

where D_{1c} is the critical damage for the uniaxial stress state and can be measured in a tensile test, σ_u is the ultimate tensile stress and

$$R_v = \left[(2/3)(1+v) + 3(1-2v)\left(\sigma_H/\sigma_{eq}\right)^2\right]$$

is called triaxiality factor, which accounts for the difference between the actual stress state and the perfectly uniaxial stress state.

This model was later implemented in the Abaqus/Explicit solver using the VUMAT subroutine [20] following the numerical algorithm proposed by Lee and Pourboghrat [21].

EXPERIMENTAL PROCEDURE

In order to determine mechanical properties and damage parameters, standard tensile tests were carried out for specimens of SAE 1050 steel for the lamellar and for the spheroidized microstructures. Three specimens for the hot rolled ferrite-pearlite material and nine specimens for the spheroidized material were tested.

The spheroidized material has been cold rolled, with a thickness reduction of 50% and subsequently annealed at 700°C for 13 hours in a 100% H_2 atmosphere, to obtain the characteristic spheroidized microstructure. The specimens were machined from a 1.0 mm thickness sheet (spheroidized material) and from a 2.0 mm thickness sheet (hot rolled material. The neck section had 75 mm length and 12.5 mm width, as shown in Figure 1.

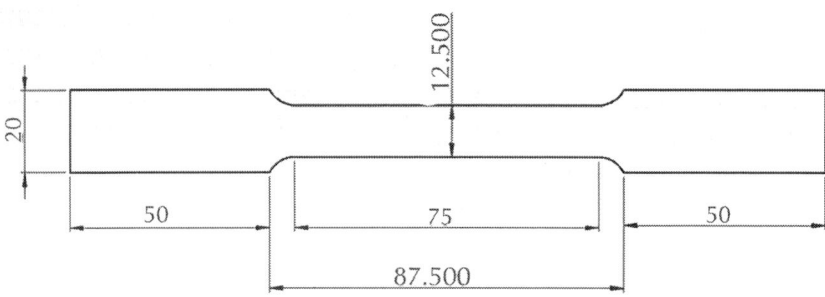

Figure 1: Workpiece dimensions, in mm, for the tensile test.

The damage variables were calculated using the variation of the elastic modulus, so several loading-unloading cycles were needed in order to measure this property with strain increase. The tests were performed in an Instron 3369 universal testing machine, with a 50 kN load cell. Each cycle began with a 1 mm crosshead displacement followed by an unloading until the force attained 50 N. The crosshead velocity was fixed at 2 mm/min. The strains were measured through a clip gage extensometer with 50 mm gage length. Sampling frequency was 5 Hz. Figure 2 illustrates the experimental setup.

Figure 3 shows the true stress-strain curves for both types of tested specimens. The loading-unloading cycles shown were used in the evaluation of the elastic modulus, measured always during the unloading path, following recommendations by Lemaitre [10]. The drop in the true stress, as pictured in this figure, can be linked to the fracture initiation. To represent the work hardening behavior of the material, Ludwik equation [22] has been used, as presented by Equation (8). The material constants are given in Table 1.

$$\sigma_Y\left(\varepsilon^p\right) = \sigma_0 + K\left(\varepsilon^p\right)^n \tag{8}$$

The evolution of the elastic modulus is shown in Figure 4 for the tested materials (including the pure iron results [23] and other carbon steels [24,25] obtained from the literature). It may be observed that the elastic modulus decreases with increasing carbon level. Also, there is a significant non-linear drop in the elastic modulus for small strains, followed by a linear evolution. Lemaitre's model considers that the damage does not occur for a strain below the critical value, and will grow with a constant rate after that value. For this reason, following the same procedure of Celentano et al. [24], any elastic modulus degradation below the linear part of the curve will be neglected. This transition coincides with the transition of the elastic regime to the plastic behavior of the material. Therefore, yielding strain will be considered as the damage strain threshold and the elastic modulus, at this point, will be assumed to be the one for the undamaged material.

The damage evolution, measured trough Equation (3), is shown in Figure 5 for both studied alloys. Critical damage D_{Ic} is taken as the damage value prior to the non-linear increase in damage, just before fracture. The other parameter to be evaluated is the damage resistance S, calculated trough Equation (9), which is obtained by manipulating Equation (6) and assuming that in the tensile test the material is under a perfectly uniaxial stress state.

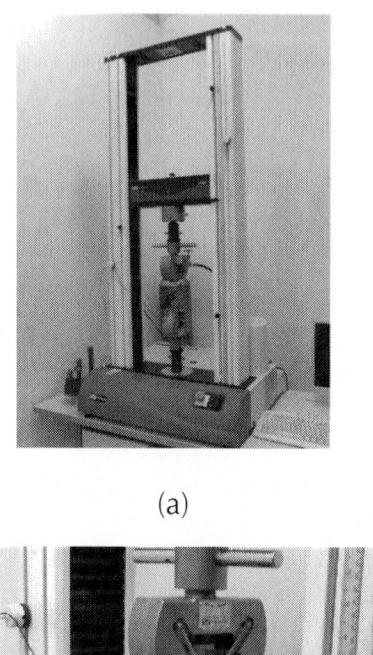

(a)

(b)

Figure 2: Experimental setup for the tensile tests.

$$\frac{\sigma_{eq}}{1-D} \leq \sigma_Y\left(\varepsilon^p\right)$$

(9)

To obtain the value of S, several experimental points must be taken from Figures 3 and 5 for different strains.

Table 1 summarizes the mechanical and damage parameters identified for SAE 1050 steel for both microstructural conditions. It must be pointed out that these parameters can be used as inputs for finite element simulations of the tensile test.

NUMERICAL SIMULATIONS

Lemaitre's model was implemented in Abaqus/Explicit finite element solver using a VUMAT subroutine aiming at the coupling of isotropic plasticity with damage, based on the stress integration algorithm called operator-split.

For each time step, the incremental strain was considered as being fully elastic, and then the corresponding stress tensor was evaluated. The von Mises criterion, coupled with damage, was used to determine if the material is indeed below the yielding condition:

$$\frac{\sigma_{eq}}{1-D} \leq \sigma_Y\left(\varepsilon^p\right)$$

(9)

If Equation (10) is not satisfied, then a plastic correcting procedure must be used to calculate the plastic increment and ensure the consistency condition. Details of this plastic corrector can be found in Lee and Pourboghrat [22]. After this calculation, stresses, the damage variable and the plastic strain are updated for the next step. Further details may be obtained in Tsiloufas [26].

The tensile test was simulated using an imposed longitudinal displacement on the right end of the specimen, with the same 2 mm/min velocity as for the experimental procedure. Boundary conditions of restricted transversal and normal displacement were imposed for both ends of the specimen, simulating the jaws of the tensile test equipment. The mesh in the test region is formed by 4500 solid hexahedral elements, with 8 nodes, linear integration and 0.5 mm length.

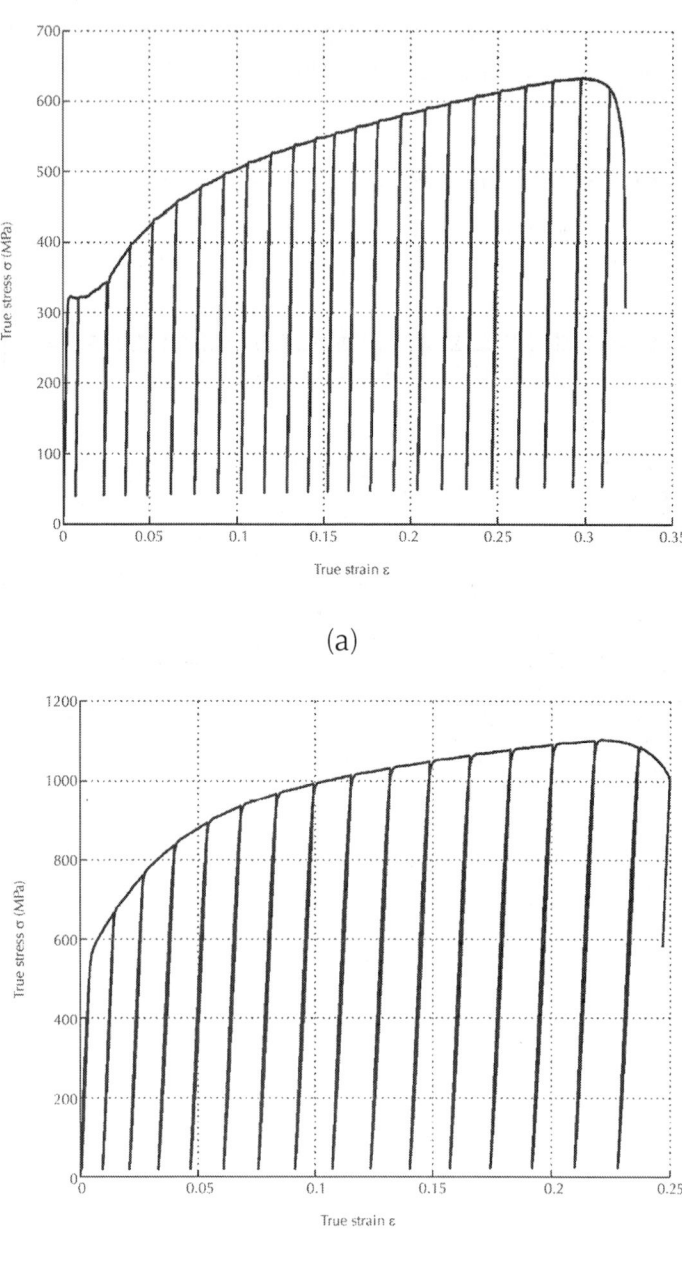

Figure 3: True stress-strain curves for SAE 1050 steel. (a) Spheroidized alloy, (b) Hot rolled alloy.

Figure 6 shows the resulting true stress-strain curves. The numerical simulations could properly reproduce the work hardening behavior of the tested materials, however the difference between the effective stress $\widetilde{\sigma_{eq}} = \sigma_{eq}/(1-D)$, that is measured through the load cell in the experiments, and the real stress σ_{eq}, must be pointed out. This may be explained by observing the material work hardening curve in terms of the real stress reaching saturation, due to the damage increase, which diminishes the material's area of resistance.

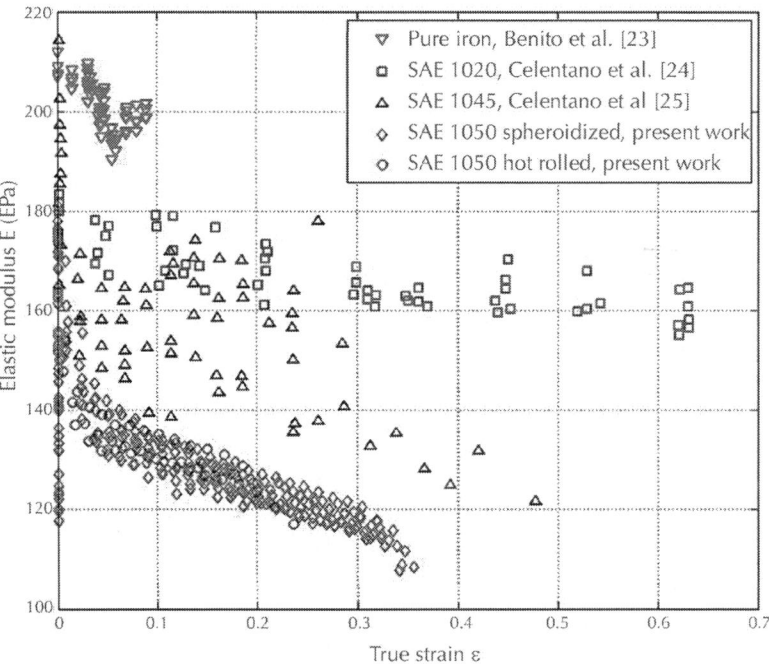

Figure 4: Elastic modulus evolution for SAE 1050 steel and comparison with pure iron and other carbon steels [23-25].

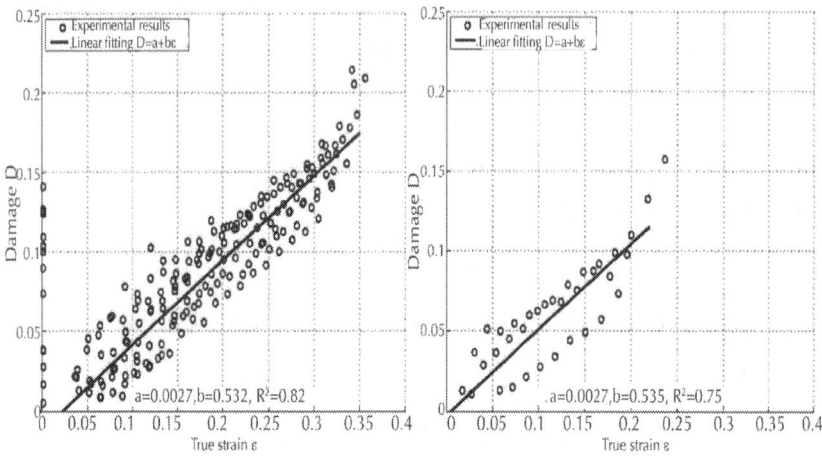

Figure 5: Ductile damage evolution versus true strain for SAE 1050. (a) Spheroidized alloy, (b) Hot rolled alloy.

Table 1: Mechanical and damage parameters for SAE 1050 steel

	Spheroidized	**Hot rolled**
E (GPa)	138	139
$V(-)$	0.29	1.29
σ_o (MPa)	290	506
σ_u (MPa)	595	1073
K (MPa)	520	998
$N(-)$	0.41	0.33
S (MPa)	1.97	4.79
ε_D (-)	0.022	0.005
$D_{Ic}(-)$	0.19	0.13

In more detail, damage evolution is shown in Figure 7. It may be observed that damage does not evolve in a linear way, as expected by Lemaitre's theory. This can be explained observing Figure 8, which shows the evolution of the triaxiality factor R_v. Lemaitre's model considers that the stress state in a tensile test is perfectly uniaxial, which is not true, as the value of R_v grows for strains higher than 0.15.

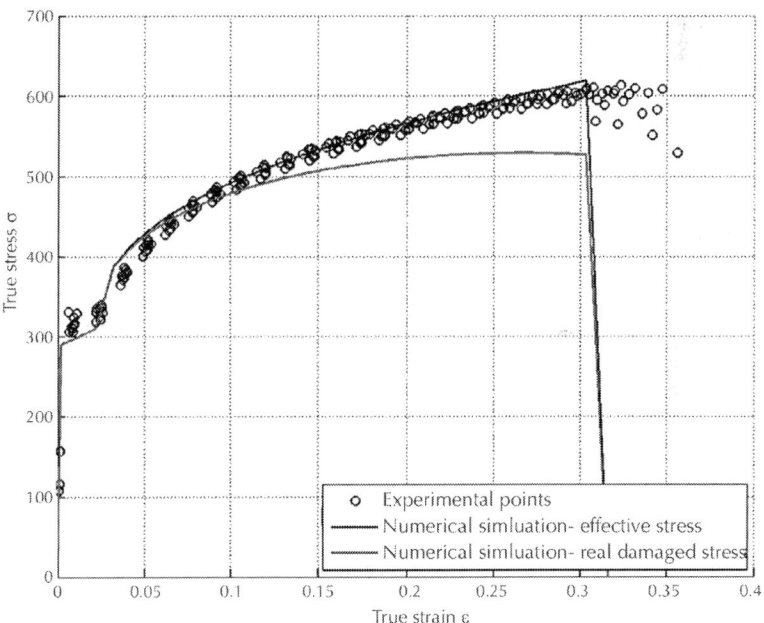

(a)

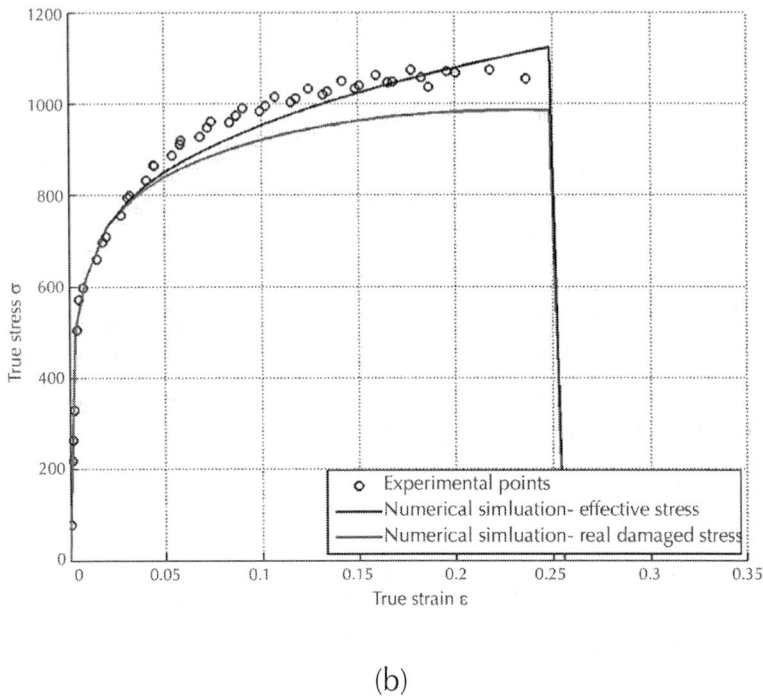

(b)

Figure 6: Strain hardening curves for SAE 1050. Experimental and numerical results. (a) Spheroidized alloy, (b) Hot rolled alloy.

Figures 9 and 10 shows the damage contour evolution for both alloys, along with a picture of the fractured experimental specimen. The fracture in the experimental test does not occur in the middle part of the sample, because any imperfection in the manufacturing of the workpiece or in the setup of the experiment may change the fracture position. It may be observed that damage is localized in the region near to the fracture, which is expected, since the localization of the strain occurring in the necking zone is the main responsible for the damage accumulation.

DAMAGE BEHAVIOR FOR CARBON STEEL ALLOYS

Using data from Refs. [24,25], Figure 11 summarizes the damage evolution in a tensile test for SAE 1020 and for 1045 steels and those of the present research. It is possible to notice that the strain threshold for damage initiation decreases and the damage growth increases with increasing carbon levels. The reason behind this fact can be associated to the microvoid formation in ductile fracture, which nucleates and grows at the interface between the ferrite matrix and the harder second phase particles [27], whose amount is proportional to the carbon level. Also, the microstructure seems to act mostly on the value of the critical damage, since for both tested SAE 1050 steels, damage initiates and increases in a similar fashion, although they present different critical damage values.

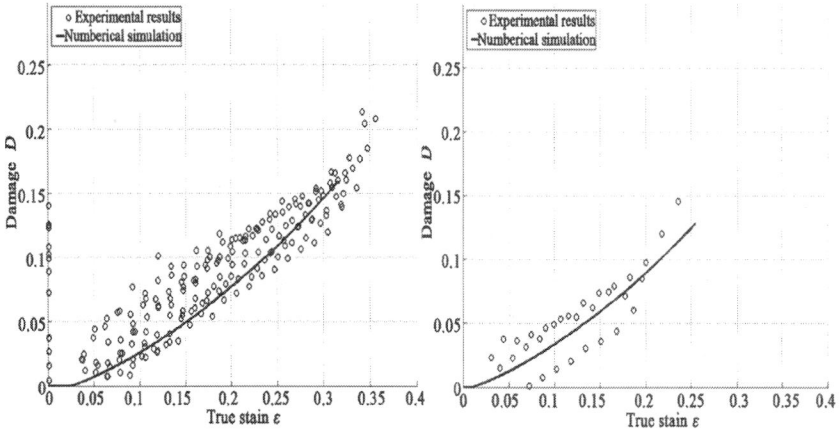

Figure 7: Damage evolution with strain for SAE 1050. Experimental and numerical results. (a) Spheroidized alloy, (b) Hot rolled alloy.

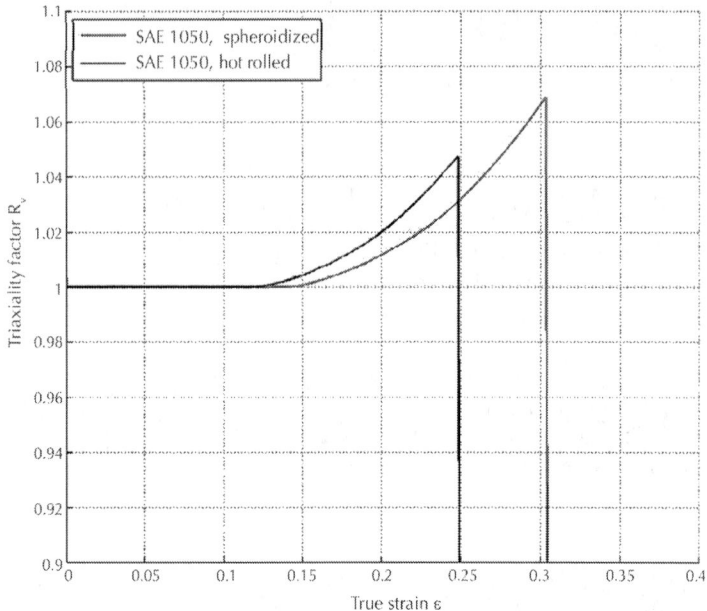

Figure 8: Evolution of the triaxiality factor for both tested materials.

For practical/industrial purposes, it is interesting to suggest a curve fitting for both damage resistance S and the damage strain threshold ε_D, as a function of the carbon level for SAE 10XX steels.

Results are shown in Figure 12. A power law of the form $y = ax^b + c$, with constants given in Table 2, has been chosen because of good agreement with experimental points.

These curves can be used as guidelines for numerical simulations in industry as a first approximation in a basic project. The uniaxial critical damage D_{1c} can be calculated using the empiric relationship of Lemaitre [10]:

(a)

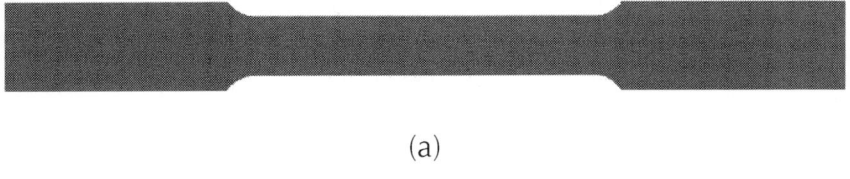

(b)

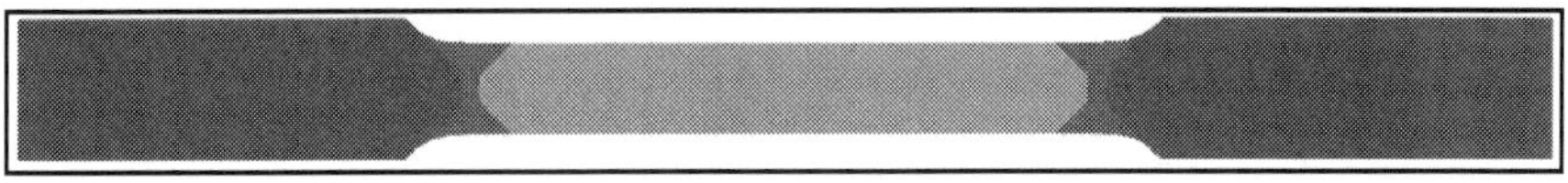

(c)

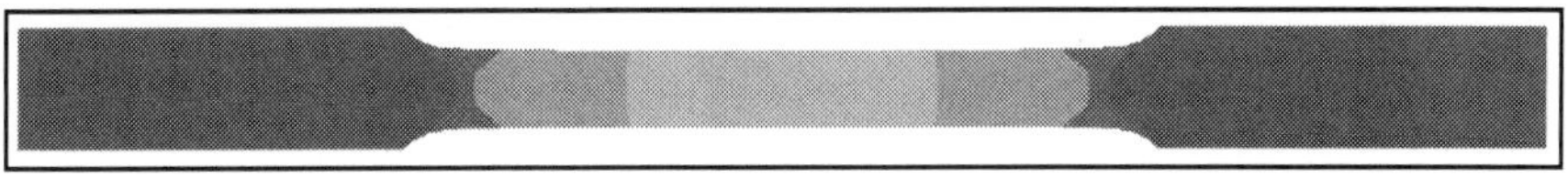

(d)

(e)

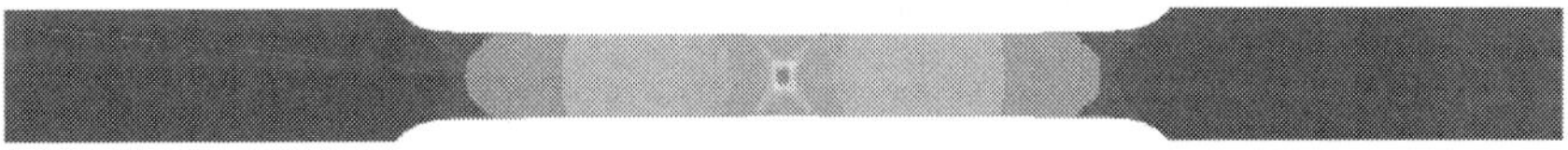

(f)

(g)

Figure 9: (a)-(f) Damage contour evolution (g) fractured experimental specimen. Spheroidized SAE 1050 steel.

(a)

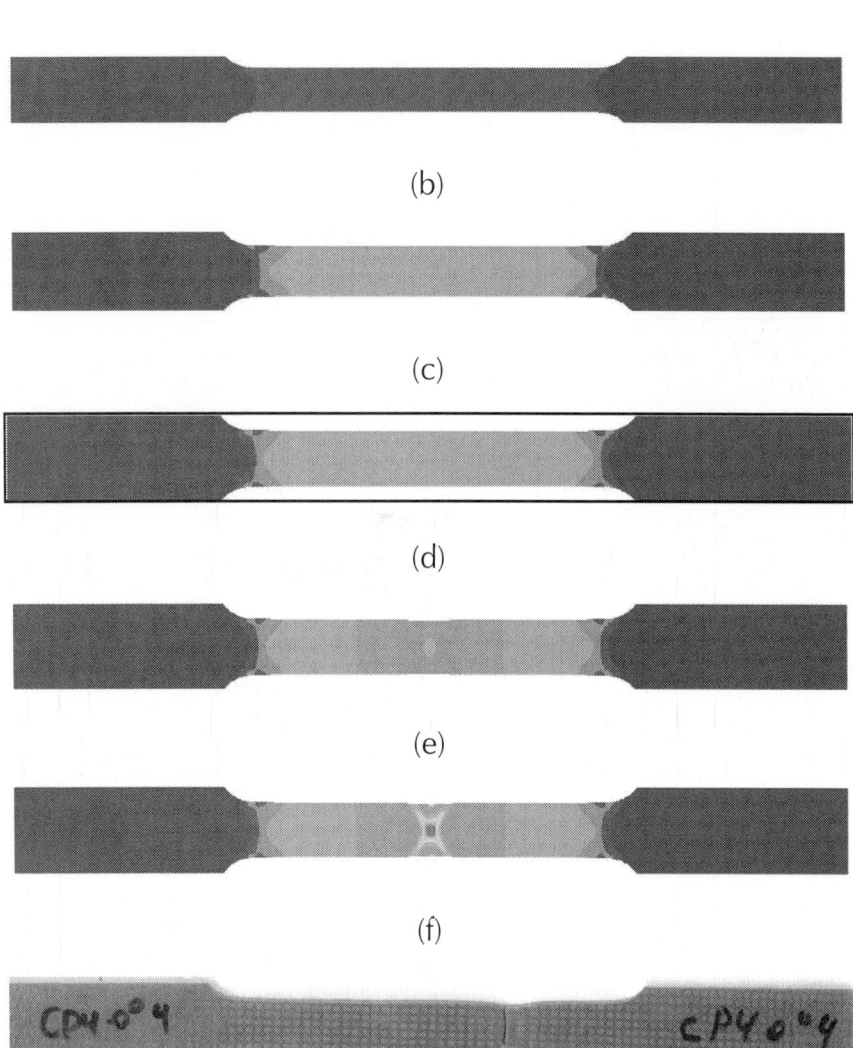

(b)

(c)

(d)

(e)

(f)

(g)

Figure 10: (a)-(f) Damage contour evolution (g) fractured experimental specimen. Hot rolled SAE 1050 steel.

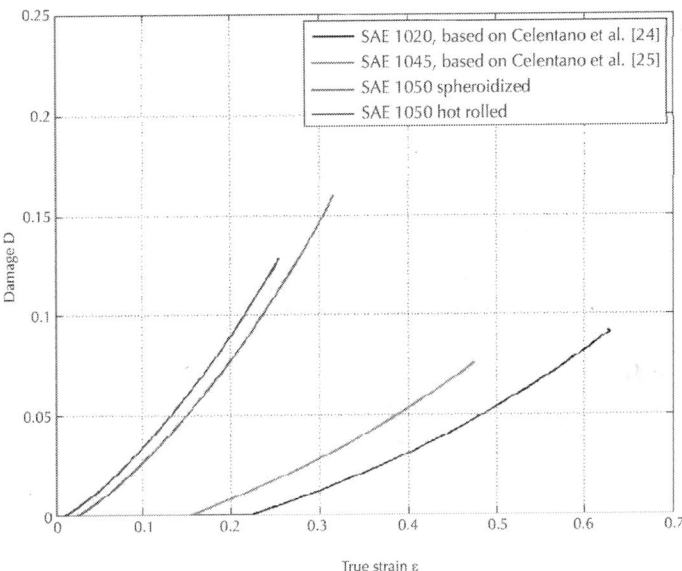

Figure 11: Damage evolution for carbon steel alloys.

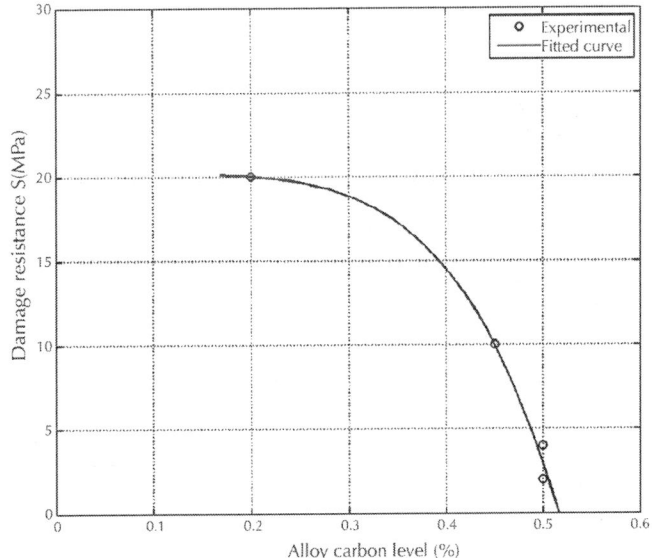

(a)

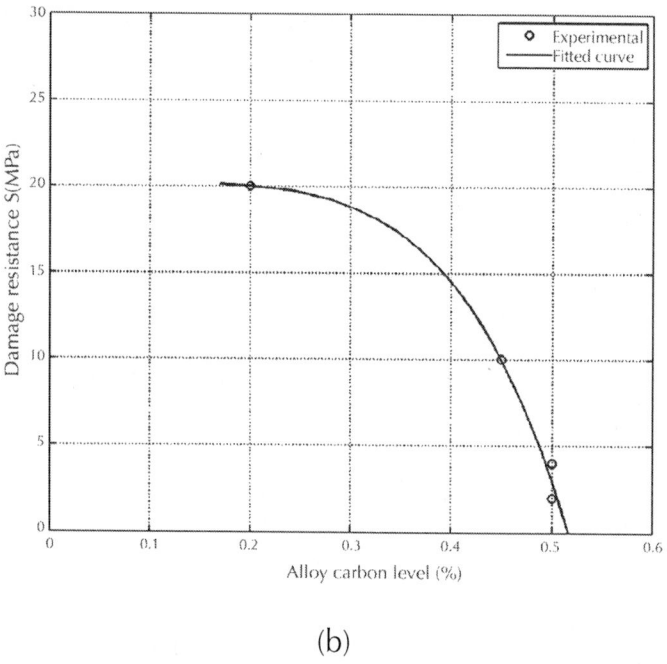

(b)

Figure 12: Power law fitting for damage properties as a function of the carbon level in SAE 10XX steels. (a) Damage resistance, (b) Damage strain threshold.

Table 2: Power law constants for damage parameters as a function of carbon level for SAE 10XX steels

Property	a	b	c
Damage resistance S (MPa)	−537.2	4.97	20.2
Damage strain threshold e_D	−252.5	10.3	0.220

$$D_{1c} = 1 - \frac{s_f}{s_u}$$

(10)

where s_f is the engineering fracture stress and s_u is the ultimate tensile engineering stress, both parameters easily provided by steel manufacturers. A comparison between the estimated and the measured value for uniaxial critical damage is shown in Table 3.

Although these guidelines can be useful, a parameter characterization procedure similar to the one here presented is highly recommended, for a more precise result.

CONCLUSIONS

In the present work, a continuum damage characterization study using loading-unloading cycles during tensile testing was performed for SAE 1050 steel for two microstructural conditions: heat-treated spheroidized cementite and lamellar ferrite-pearlite (hot rolled). Damage itself was indirectly measured by means of the degradation of the elastic modulus. Mechanical parameters for modeling the work hardening, behavior and the damage evolution were evaluated for the Lemaitre's ductile damage model.

Table 3: Measured and estimated uniaxial critical damage for SAE 1050 steel

	Spheroidized	Hot rolled
Ultimate engineering stress s_u (MPa)	480	790
Fracture engineering stress s_f(MPa)	400	900
Estimated critical damage $D_{lc'}$	0.17	0.12
Calculated critical damage $D_{lc'}$	0.19	0.13
Error in critical damage estimation	10.5%	7.7%

A numerical algorithm was written to account for the coupling between damage and isotropic plasticity, and implemented in Abaqus/Explicit solver by means of a VUMAT subroutine. Then, simulations of the tensile test were performed, providing good agreement with experimental results. The difference between effective stresses and the real stresses acting on a damaged volume element is presented. Numerical damage evolution was not linear with strain, as would be expected. The explanation lies on the stress-state, which ceases to be perfectly uniaxial for a strain higher than about 0.15.

Although damage mechanics cannot describe macrocrack formation that takes place just prior to fracture, using such a model together with experimental characterization procedure seem to be a useful manner for predicting the initial stages of a ductile failure phenomenon.

ACKNOWLEDGEMENTS

The authors would like to thanks Brasmetal Waeholtz for providing the material samples for the tests, GMSIEPOLI/USP for the tensile test machine and CAPES for the scholarship of SPT, provided for the development of this study.

REFERENCES

1.   M. G. Cockroft and D. J. Latham, "Ductility and the Workability of Metals," Journal of the Institute of Metals, Vol. 96, 1968, pp. 33-39.

2.   M. Oyane, T. Sato, K. Okimoto and S. Shima, "Criteria for Ductile Fracture and Their Applications," Journal of Mechanical Working Technology, Vol. 4, No. 1, 1980, pp. 65-81.doi:10.1016/0378-3804(80)90006-6

3.   A. L. Gurson, "Continuum Theory of Ductile Rupture by Void Nucleation and Growth: Part I—Yield Criteria and Flow Rules for Porous Ductile Media," Journal of Engineering Materials and Technology, Vol. 99, No. 1, 1977, pp. 2-15. doi:10.1115/1.3443401

4.   L. M. Kachanov, "Rupture Time under Creep Conditions," International Journal of Fracture, Vol. 97, 1999, pp. 11-18.

5.   J. L. Chaboche, "Continuum Damage Mechanics: Present State and Future Trends," Nuclear Engineering and Design, Vol. 105, No. 1, 1987, pp. 19-33. doi:10.1016/0029-5493(87)90225-1

6.   J. Lemaitre, "How to Use Damage Mechanics," Nuclear Engineering and Design, Vol. 80, No. 1, 1984, pp. 233- 245. doi:10.1016/0029-5493(84)90169-9

7.   J. Lemaitre, "A Continuous Damage Mechanics Model for Ductile Fracture," Journal of Engineering Materials and Technology, Vol. 77, 1985, pp. 335-344.

8.   J. L. Chaboche, "Continuum Damage Mechanics: Part I—General Concepts," Journal of Applied Mechanics, Vol. 55, No. 1, 1988, pp. 55-64. doi:10.1115/1.3173661

9.   J. L. Chaboche, "Continuum Damage Mechanics: Part II—Damage Growth, Crack Initiation and Crack Growth," Journal of Applied Mechanics, Vol. 55, No. 1, 1988, pp. 65-72. doi:10.1115/1.3173662

10. J. Lemaitre, "A Course on Damage Mechanics," 2nd Edition, Springer, Berlin, 1996.doi:10.1007/978-3-642-18255-6

11. C. L. Chow and J. Wang, "An Anisotropic Theory of Continuum Damage Mechanics for Ductile Fracture," Engineering Fracture Mechanics, Vol. 27, No. 5, 1987, pp. 547-558. doi:10.1016/0013-7944(87)90108-1

12. W. Tai and B. Yang, "A New Damage Mechanics Criterion for Ductile Fracture," Engineering Fracture Mechanics, Vol. 27, No. 4, 1987, pp. 371-378. doi:10.1016/0013-7944(87)90174-3

13. T.-J. Wang, "Unified CDM Model and Local Criterion for Ductile Fracture: I—Unified CDM Model for Ductile Fracture," Engineering Fracture Mechanics, Vol. 42, 1992, pp. 177-183.

14. T.-J. Wang, "Unified CDM Model and Local Criterion for Ductile Fracture: II—Ductile Fracture Local Criterion Based on the CDM Model," Engineering Fracture Mechanics, Vol. 42, 1992, pp. 185-193.

15. S. Chandrakanth and P. Pandey, "An Isotropic Damage Model for Ductile Material," Engineering Fracture Mechanics, Vol. 50, No. 4, 1995, pp. 457-465. doi:10.1016/0013-7944(94)00214-3

16. N. Bonora, "A Nonlinear CDM Model for Ductile Failure," Engineering Fracture Mechanics, Vol. 58, No. 1, 1997, pp. 11-28. doi:10.1016/S0013-7944(97)00074-X

17. L. Storojeva, D. Ponge, R. Kaspar and D. Raabe, "Development of Microstructure and Texture of Medium Carbon Steel during Heavy Warm Deformation," Acta Materialia, Vol. 52, No. 8, 2004, pp. 2209-2220. doi:10.1016/j.actamat.2004.01.024

18. F. A. McClintock, "A Criterion for Ductile Fracture by the Growth of Holes," Journal of Applied Mechanics, Vol. 35, No. 2, 1968, pp. 363-371. doi:10.1115/1.3601204

19. F. Sidoroff, "On the Formulation of Plasticity and Viscoplasticity with Internal Variables," Archiwum Mechaniki Stosowanej, 1975, pp. 807-819.

20. Simulia Abaqus 6.10, "User Subroutines Reference Manual," 2010.

21. S. W. Lee and F. Pourboghrat, "Finite Element Simulation of the Punchless Piercing Process with Lemaitre Damage Model," International Journal of Mechanical Sciences, Vol. 47, No. 11, 2005, pp. 1756-1768. doi:10.1016/j.ijmecsci.2005.06.009

22. P. Ludwik, "Elements der Technologischen Mechanik," 3rd Edition, Vol. 32, Springer, Berlin, 1909.

23. J. A. Benito, J. M. Manero, J. Jorba and A. Roca, "Change of Young's Modulus of Cold-Deformed Pure Iron in a Tensile Test," Metallurgical and Materials Transactions A, Vol. 36A, 2005, pp. 3317-3324.

24. D. J. Celentano and J.-L. Chaboche, "Experimental and Numerical Characterization of Damage Evolution in Steels," International Journal of Plasticity, Vol. 23, No. 10-11, 2007, pp. 1739-1762. doi:10.1016/j.ijplas.2007.03.008

25. D. J. Celentano, P. E. Tapia and J. L. Chaboche, "Experimental and Numerical Characterization If Damage Evolution in

Steels," In: G. Buscaglia, E. Dari and O. Zamonsky, Eds., Mecánica Computacional, Vol. XXIII, Bariloche, Argentina, 2004, pp. 45-58.

26. S. P. Tsiloufas, "Estudo da Fratura Dúctil em Chapas de Médio Carbono Sob a Ótica da Teoria da Mecânica do Dano (in Portuguese)," MSc. Dissertation, University of São Paulo, São Paulo, 2012.

27. J. Gurland, "Observations on the Fracture of Cementite Particles in a Spheroidized 1.05% C Steel Deformed at Room Temperature," Acta Metallurgica, Vol. 20, No. 5, 1972, pp. 735-741. doi:10.1016/0001-6160(72)90102-2

Citations

CHAPTER 1

Cristina Alia, María V. Biezma, Paz Pinilla, José M. Arenas, and Juan C. Suárez, "Degradation in Seawater of Structural Adhesives for Hybrid Fibre-Metal Laminated Materials," Advances in Materials Science and Engineering, vol. 2013, Article ID 869075, 10 pages, 2013. doi:10.1155/2013/869075.

CHAPTER 2

X. W. Ye, Y. H. Su, and J. P. Han, "Structural Health Monitoring of Civil Infrastructure Using Optical Fiber Sensing Technology: A

Index